Lecture Notes in Business Information Processing 413

Series Editors

More information about this series at http://www.springer.com/series/7911

Ewa Ziemba · Witold Chmielarz (Eds.)

Information Technology for Management

Towards Business Excellence

15th Conference, ISM 2020
and FedCSIS-IST 2020 Track, Held as Part of FedCSIS
Sofia, Bulgaria, September 6–9, 2020
Extended and Revised Selected Papers

Editors
Ewa Ziemba
University of Economics in Katowice
Katowice, Poland

Witold Chmielarz
University of Warsaw
Warsaw, Poland

ISSN 1865-1348 ISSN 1865-1356 (electronic)
Lecture Notes in Business Information Processing
ISBN 978-3-030-71845-9 ISBN 978-3-030-71846-6 (eBook)
https://doi.org/10.1007/978-3-030-71846-6

This Springer imprint is published by the registered company Springer Nature Switzerland AG
The registered company address is: Gewerbestrasse 11, 6330 Cham, Switzerland

Preface

Five editions of this book appeared in the last five years:

- *Information Technology for Management* in 2016 (LNBIP 243);
- *Information Technology for Management: New Ideas or Real Solutions* in 2017 (LNBIP 277);
- *Information Technology for Management: Ongoing Research and Development* in 2018 (LNBIP 311);
- *Information Technology for Management: Emerging Research and Applications* in 2019 (LNBIP 346); and
- *Information Technology for Management: Current Research and Future Directions* in 2020 (LNBIP 380).

Given the rapid developments in information systems and technologies, and their adoption for improving business and public organizations, there was a clear need for an updated version.

The present book includes extended and revised versions of a set of selected papers submitted to the Information Systems and Technologies conference track (FedCSIS-IST 2020) organized within the 15th Conference on Computer Science and Information Systems (FedCSIS 2020), held in Sofia, Bulgaria, during September 6–9, 2020.

FedCSIS provides a forum for bringing together researchers, practitioners, and academics to present and discuss ideas, challenges, and potential solutions on established or emerging topics related to research and practice in computer science and information systems. Since 2012, Proceedings of FedCSIS are indexed in the Thomson Reuters Web of Science, Scopus, IEEE Xplore Digital Library, and other indexing services.

FedCSIS-IST covers a broad spectrum of topics which bring together the sciences of information technologies, information systems, and social sciences, i.e., economics, management, business, finance, and education. This edition consisted of the following four scientific sessions: Advances in Information Systems and Technologies (AIST 2020), 2nd Special Session on Data Science in Health, Ecology and Commerce (DSH 2020), 15th Conference on Information Systems Management (ISM 2020), and 26th Conference on Knowledge Acquisition and Management (KAM 2020).

AIST seeks the most recent innovations, current trends, professional experiences, and new challenges in the several perspectives of information systems and technologies, i.e., design, implementation, stabilization, continuous improvement, and transformation. IT covers business intelligence, big data, data mining, machine learning, cloud computing, mobile applications, social networks, internet of things, sustainable technologies and systems, blockchain, etc.

DSH focuses on all forms of data analysis, data economics, information systems, and data-based research, focusing on the interaction of the four fields, i.e., health, ecology, and commerce.

ISM concentrates on various issues of planning, organizing, resourcing, coordinating, controlling, and leading management functions to ensure the smooth operation of information systems in organizations.

KAM discusses approaches, techniques, and tools in knowledge acquisition and other knowledge management areas with focus on the contribution of artificial intelligence for improvement of human-machine intelligence and facing the challenges of this century.

For FedCSIS-IST 2020, we received 29 papers from 18 countries on six continents. The quality of the papers was evaluated by the members of the Program Committees by taking into account the criteria for papers' relevance to conference topics, originality, and novelty. After extensive reviews, only 6 papers were accepted as full papers and 6 as short papers. Finally, 8 papers of the highest quality were carefully reviewed and chosen by the Chairs of the four sessions, and the authors were invited to extend their research and submit the new extended papers for consideration for this LNBIP publication. Our guiding criteria for including papers in the book were the excellence of the papers as indicated by the reviewers, the relevance of the subject matter for improving management by adopting information technology, as well as the promise of the scientific contributions and the implications for practitioners. The selected papers reflect state-of-art research work that is often oriented toward real-world applications and highlight the benefits of information systems and technologies for business and public administration, thus forming a bridge between theory and practice.

The papers selected to be included in this book contribute to the understanding of relevant trends of current research on and future directions of information systems and technologies for achieving business excellence. The first part of the book focuses on improving information systems project management methods, the second part presents numerical methods for solving management problems, and the third part explores technological infrastructure for business excellence.

Finally, we and the authors hope readers will find the content of this book useful and interesting for their own research activities. It is in this spirit and conviction we offer our monograph, which is the result of the intellectual effort of the authors, for the final judgment of the readers. We are open to discussion on the issues raised in this book, as well as looking forward to critical or even polemical voices as to the content and form.

February 2021

Ewa Ziemba
Witold Chmielarz

Acknowledgment

We would like to express our gratitude to all those people who helped create the success of the FedCSIS-IST research events. First of all, we want to thank the authors for offering their very interesting research and submitting their new findings for publication in LNBIP. We express our appreciation to the members of the Program Committees for taking the time and effort necessary to provide valuable insights for the authors. The high standards followed by them enabled the authors to ensure the high quality of their papers. Their work enabled us to ensure the high quality of the conference sessions, excellent presentations of the authors' research, and valuable scientific discussion. We acknowledge the Chairs of the FedCSIS 2020, i.e., Maria Ganzha, Leszek A. Maciaszek, and Marcin Paprzycki, for building an active international community around the FedCSIS conference. Last but not least, we are indebted to the Springer-Verlag team headed by Ralf Gerstner and Alfred Hofmann without whom this book would not have been possible. Many thanks also to Christine Reiss and Raghuram Balasubramanian for handling the production of this book.

Organization

FedCSIS-IST 2020

Chairs

Ewa Ziemba	University of Economics in Katowice, Poland
Guangming Cao	Ajman University, United Arab Emirates
Daphne Raban	University of Haifa, Israel

AIST 2020

Chairs

Ewa Ziemba	University of Economics in Katowice, Poland
Alberto Cano	Virginia Commonwealth University, USA
Jarosław Wątróbski	University of Szczecin, Poland

Program Committee

Ofir Ben-Assuli	Ono Academic College, Israel
Dizza Beimel	Ruppin Academic Center, Israel
Andrzej Białas	Institute of Innovative Technologies EMAG, Poland
Witold Chmielarz	University of Warsaw, Poland
Dimitar Christozov	American University in Bulgaria, Bulgaria
Krzysztof Cios	Virginia Commonwealth University, USA
Pankaj Deshwal	Netaji Subhas University of Technology, India
Gonçalo Paiva Dias	Universidade de Aveiro, Portugal
Krzysztof Kania	University of Economics in Katowice, Poland
Agnieszka Konys	West Pomeranian University of Technology in Szczecin, Poland
Eugenia Kovacheva	University of Library Studies and Information Technologies, Bulgaria
José Luna	University of Córdoba, Spain
Krzysztof Michalik	University of Economics in Katowice, Poland
Zbigniew Pastuszak	Maria Curie-Skłodowska University, Poland
Daphne Raban	University of Haifa, Israel
Elisabeth Rakus-Andersson	Blekinge Institute of Technology, Sweden
Amit Rechavi	Ruppin Academic Center, Israel
Nina Rizun	Gdańsk University of Technology, Poland
Kamel Rouibah	College of Business Administration, Kuwait University, Kuwait
Yonit Rusho	Shenkar College, Israel
Joanna Santiago	ISEG - University of Lisbon, Portugal

Wojciech Sałabun	West Pomeranian University of Technology in Szczecin, Poland
Marcin Sikorski	Gdańsk University of Technology, Poland
Jacek Szołtysek	University of Economics in Katowice, Poland
Łukasz Tomczyk	Pedagogical University of Cracow, Poland
Bob Travica	University of Manitoba, Canada
Isaac Triguero Velázquez	University of Nottingham, UK
Jarosław Wątróbski	University of Szczecin, Poland
Paweł Ziemba	University of Szczecin, Poland

DSH 2020

Chairs

Bogdan Franczyk	University of Leipzig, Germany
Carsta Militzer-Horstmann	WIG2 Institute for Health Economics and Health Service Research, Germany
Dennis Häckl	WIG2 Institute for Health Economics and Health Service Research, Germany
Jan Bumberger	Helmholtz-Centre for Environmental Research – UFZ, Germany
Olaf Reinhold	University of Leipzig/Social CRM Research Center, Germany

Program Committee

Adil Alpkoçak	Dokuz Eylul University, Turkey
Douglas da Rocha Cirqueira	Dublin City University, Ireland
Nilanjan Dey	Techno India College of Technology, India
Nils Kossack	WIG2 Institute for Health Economics and Health Service Research, Germany
Karol Kozak	Fraunhofer and Uniklinikum Dresden, Germany
Piotr Popowski	Medical University of Gdańsk, Poland
Shelly Sachdeva	National Institute of Technology Delhi, India
Katarzyna Wasielewska-Michniewska	Systems Research Institute of the Polish Academy of Sciences, Poland
Danny Wende	WIG2 Institute for Health Economics and Health Service Research/Technical University Dresden, Germany

ISM 2020

Chairs

Bernard Arogyaswamy	Le Moyne University, USA
Witold Chmielarz	University of Warsaw, Poland
Jarosław Jankowski	West Pomeranian University of Technology in Szczecin, Poland

Dimitris Karagiannis	University of Vienna, Austria
Jerzy Kisielnicki	University of Warsaw, Poland
Ewa Ziemba	University of Economics in Katowice, Poland

Program Committee

Mohammed Alonazi	University of Sussex, UK
Janis Bicevskis	University of Latvia, Latvia
Boyan Bontchev	Sofia University St Kliment Ohridski, Bulgaria
Bolesław Borkowski	Warsaw University of Life Sciences, Poland
Alberto Cano	Virginia Commonwealth University, USA
Vincenza Carchiolo	University of Catania, Italy
Beata Czarnacka-Chrobot	Warsaw School of Economics, Poland
Robertas Damaševičius	Silesian University of Technology, Poland
Pankaj Deshwal	Netaji Subhas University of Technology, India
Yanqing Duan	University of Bedfordshire, UK
Monika Eisenbardt	University of Economics in Katowice, Poland
Ibrahim El Emary	King Abdulaziz University, Saudi Arabia
Susana de Juana Espinosa	University of Alicante, Spain
Marcelo Fantinato	University of São Paulo, Brazil
Renata Gabryelczyk	University of Warsaw, Poland
Nitza Geri	The Open University of Israel, Israel
Krzysztof Kania	University of Economics in Katowice, Poland
Andrzej Kobyliński	Warsaw School of Economics, Poland
Christian Leyh	Dresden University of Technology, Germany
Michele Malgeri	University of Catania, DIEEI, Italy
Oana Merkt	University of Hohenheim, Germany
Karolina Muszyńska	University of Szczecin, Poland
Walter Nuninger	Polytech Lille/Université de Lille, France
Nina Rizun	Gdańsk University of Technology, Poland
Uldis Rozevskis	University of Latvia, Latvia
Andrzej Sobczak	Warsaw School of Economics, Poland
Jakub Swacha	University of Szczecin, Poland
Symeon Symeonidis	Democritus University of Thrace, Greece
Edward Szczerbicki	University of Newcastle, Australia
Oskar Szumski	University of Warsaw, Poland
Bob Travica	University of Manitoba, Canada
Jarosław Wątróbski	University of Szczecin, Poland
Janusz Wielki	Opole University of Technology, Poland
Dmitry Zaitsev	Odessa State Environmental University, Ukraine

KAM 2020

Chairs

Krzysztof Hauke	Wroclaw University of Economics, Poland
Małgorzata Nycz	Wroclaw University of Economics, Poland
Mieczysław Owoc	Wroclaw University of Economics, Poland
Maciej Pondel	Wroclaw University of Economics, Poland

Program Committee

Witold Abramowicz	Poznań University of Economics and Business, Poland
Frederic Andres	National Institute of Informatics, Japan
Yevgeniy Bodyanskiy	Kharkiv National University of Radio Electronics, Ukraine
Witold Chmielarz	University of Warsaw, Poland
Dimitar Christozov	American University in Bulgaria, Bulgaria
Jan Vanthienen	Katholieke Universiteit Leuven, Belgium
Eunika Mercier-Laurent	University Jean Moulin Lyon 3, France
Małgorzata Sobińska	Wroclaw University of Economics, Poland
Jerzy Surma	Warsaw School of Economics, Poland/University of Massachusetts Lowell, USA
Julian Vasilev	University of Economics - Varna, Bulgaria
Yungang Zhu	Jilin University, China

Contents

Improving Project Management Methods

Digital Assets for Project-Based Studies and Project Management 3
Gloria J. Miller

Data Quality Model-Based Testing of Information Systems: Two-Level Testing of the Insurance System 25
Anastasija Nikiforova, Janis Bicevskis, Zane Bicevska, and Ivo Oditis

Advanced Comprehension Analysis Using Code Puzzle: Considering the Programming Thinking Ability 45
Hiroki Ito, Hiromitsu Shimakawa, and Fumiko Harada

Numerical Methods of Solving Management Problems

Wind Farms Maintenance Optimization Using a Pickup and Delivery VRP Algorithm. 67
Vincenza Carchiolo, Alessandro Longheu, Michele Malgeri, Giuseppe Mangioni, and Natalia Trapani

An Improved Map Matching Algorithm Based on Dynamic Programming Approach 87
Alexander Yumaganov, Anton Agafonov, and Vladislav Myasnikov

A Cluster-Based Approach to Solve Rich Vehicle Routing Problems 103
Emir Zunic, Sead Delalic, Dzenana Donko, and Haris Supic

Technological Infrastructure for Business Excellence

Street Addressing System as an Essential Component in Sustainable Development and Realization of Smart City Conception 127
Dmitriy Gakh

Digital Transformation in Legal Metrology: An Approach to a Distributed Architecture for Consolidating Metrological Services and Data 146
Alexander Oppermann, Samuel Eickelberg, and John Exner

Author Index 165

Improving Project Management Methods

Digital Assets for Project-Based Studies and Project Management

Gloria J. Miller DBA(✉)

maxmetrics, Heidelberg, Germany
g.j.m@ieee.org

Abstract. This research provides a literature survey of digital assets available through a project; specifically, it identifies sources of data that can be used for practicing data-driven, context-specific project management, or for project-based academic research. Projects are key vehicles for economic and social action, and they are also a primary source of innovation, research, and organizational change. The project boundaries of time, tasks, and people define a rich environment for collecting behavioral and attitudinal data for learning opportunities and academic research. Based on a systematic literature review of the top four project management journals, this research identifies four categories of data sources – communications, reports/records, model representations, and computer systems – and 52 digital assets. The list of digital assets can be inputs for the creation of project artifacts as well as sources for monitoring and controlling project activities and for sense-making in retrospectives or lessons learned. In an illustrative case, this research uses three of the digital assets, social network analysis, and topic modeling to analyze the verbal and written communications between project participants. The classification model and categorization are useful for decision support and artificial intelligence systems model development that requires real-world data.

Keywords: Digital asset · Project management · Project studies · Big data

1 Introduction

Projects offer rich environments for conducting research and learning [1, 2] and for practicing data-driven, context-specific project management [3]. They are a key vehicle for economic and social action, as well as a primary source of innovation, research, and organizational change [4–6]. They can involve budgets larger than the gross domestic product of a small nation and resources greater than the organizations participating in them [1]. Some of the factors that make projects interesting for analysis through a multitude of theoretical lenses are the scale, complexity, uncertainty, and geographic distributions [6]. Projects can be explained and studied using philosophical underpinnings (such as the Newtonian understanding of time, space, and activity) through the project archetypes such as project-based organizations, project-supported organizations, and project networks, or through the investigation of the changes in project processes or actors [4–6].

E. Ziemba and W. Chmielarz (Eds.): ISM 2020/FedCSIS-IST 2020, LNBIP 413, pp. 3–24, 2021.
https://doi.org/10.1007/978-3-030-71846-6_1

The variety and richness that make projects interesting to study, however, can make them a challenge to manage efficiently. First, there are many project-specific tools and techniques available to manage project complexities [7, 8]. Muszyńska and Swacha [8] highlighted multinational, dispersed teams, and the need for a different tool selection for each unique project as some of the factors that make projects complex. Second, the tens of project management suites and tools available demonstrate the many factors that managers must consider when planning, monitoring, and controlling projects [7–9]. Third, although collecting lessons learned and implementing improvement processes are central concepts in project management standards [10–12], the learning rarely happens or does not deliver the intended results [13]. Finally, the administration of projects is moving away from paper documents to a new way of managing task infrastructure through digital information [14].

Researchers have begun to argue that real-time project data should be used in stakeholder engagement [15], performance management [3, 16, 17], monitoring and controlling [18], and policy setting. These approaches support project management moving from individual human-based decisions to expert decisions to utilizing artificial intelligence. For example, Snider, Gopsill, Jones, Emanuel, and Hicks [3] argue that project performance should be evaluated based on an analysis of the data artifacts produced from everyday project activities rather than by relying on managerial understanding. Nemati, Todd, and Brown [19] explain that project estimation is suitable for an artificial neural network given the numerous potential project configurations. Willems and Vanhoucke [18] found that artificial intelligence was used at the front-end of projects but suggested its use has been less investigated during projects. The transition to these data-driven methods is supported by the growing importance of digital workflows and analytics in project delivery [14].

Therefore, even though projects are rich grounds for research and the push towards data-driven project management, the topic of digital data—structured and non-structured—in projects is not sufficiently covered in project management literature. This research involves a systematic literature review to compile a list of digital assets available through a project context. A digital asset classification would be valuable to project researchers and to project managers for practicing data-driven, context-specific project management, or for project-based academic research. Thus, the study uses a conceptual model that presumes digital assets are sources of learning. While there are individual studies that provide some insight into the sources of project management data and while the project management standards provide document lists, there is no comprehensive list of project-specific digital assets available in the literature. Furthermore, this study supports the call for new research approaches that investigate the actual or lived experience [2].

The paper is structured as follows. Section two provides a literature review. Section three includes the conceptual framework and a description of the research methodology. Section four defines the classification model and describes its characteristics. Section five discusses the model and section six provides the study's contribution, implications, limitations and consideration for future research.

2 Literature Review

2.1 Project Management Tools

Projects are a temporary organization with a set of actors working together over a limited, pre-determined period of time for a given activity [1]. They can vary in size and scope from organizational projects to programs used to transform society. There are hundreds of project management tools and tens of software suites available to manage the complexities created by the unique project environments. Project management tools include methods, decision-making techniques or models, risk assessment tools, information communication technology support tools, computer models, databases, indices and simulations [7–9, 20, 21]. Online collaborative toolsets allow project-related material to be customized for specific project roles and made available to multiple teams across multiple sites and countries [14]. Tool diversity, lack of use, limited feature coverage, and role distribution are some of the limitations that prevent project management tools and systems from being a single source of data for project research. Furthermore, the extent of use of the tools varies by the type of project deliverables. Finally, while databases of historical costs, lessons-learned, and risks from past projects are perceived to be of great value, they are some of the least-used tools [7].

2.2 Lessons Learned

In the available literature, organizational learnings rarely occur or deliver their intended results [13], and project management lessons learned are ineffective, incomplete, and ill-conceived [13, 22]. Consequently, there has been a call for an improvement in project management processes and methods for disseminating organizational learnings. For example, the Syllk model is useful for identifying needed improvements but not for making the underlying changes to policies, systems, and processes [13]. The triple-loop learning model combines individual, project, and organizational learning processes to react to the changes needed to embed learnings into an organization's policies, standards, and practices [22]. However, the reality is a minority of organizations identify and capture lessons learned through formalized procedures. Moreover, knowledge is subjective and dependent upon the individual. The more successful individuals are at learning and retaining the learnings, the more difficulty they have in relating the salient knowledge to their peers [22, 23]. Specifically, knowledge is a driver for organizational success [24]. Capturing knowledge through digital data is commercially important and supports the trend towards using artificial intelligence and traditional statistical methods to solve project management problems [3, 18, 25].

2.3 Digital Data in Project Management

Quinton and Reynolds [26] defined digital media as encompassing "all computer-mediated internet and digitally-enabled media through which data may be collected, shared and/or analyzed..." [26, p. 11]. They described digital data (without providing a formal definition) as data available through digital technologies. Snider, Gopsill, Jones, Emanuel, and Hicks [3] limited their definition of digital data to engineering data produced from every-day project activities. Whyte [14] determined that digital information permits the introduction of collaborative project delivery models with configurable supply-chains and relationships. In these new models, project teams are contractually obliged to deliver digital information to the long-term owner or operator. The study identified the convergence of digital information and project practices as a future area of research.

2.4 Summary

In summary, projects are important vehicles for delivering economic value. They can be complicated and can require managers to select from a variety of tools and methods for their efficient management. However, projects can also produce valuable digital data and individual and organizational learnings. While project contexts are promising for research, integrating individual learnings into organizations is difficult. For this study, we define a digital asset as any digital data that may be used for analysis using computer technology. Digital assets are an increasingly valuable way of accessing project learning. Furthermore, they are useful as a foundation for project management decision-making and artificial intelligence systems. However, there is a gap in providing a comprehensive list of digital assets available from a project context. The objective of this study is to compile a survey of digital assets available through a project-context that can be used for research and for moving towards data-driven project management.

3 Research Methodology

First, we provide a conceptual model for investigating the topic. Next, we described the systematic literature review and illustrative case methods used in the study. Finally, we explain the measures taken to ensure validity and reliability.

3.1 Conceptual Model

The conceptual model for this study builds on theories from organizational learning and concepts from the triple-loop learning model [22]. The triple-loop learning model aligns the project, process, and organizational learning perspectives with the planning, delivery, and temporal phases of a project. Each learning loop considers a targeted outcome against which project variances are evaluated. Understanding and investigating variances produces learnings. Learnings may be incorporated into processes, procedures, or other project objects based on decisions and reactions to the variance. Learning goals are defined and further project actions are taken.

The conceptual framework from this study presumes that the source of learning is the digital asset. Figure 1 demonstrates the conceptual model for the relationship between digital assets and project artifact. Digital assets are used and are generated at all stages of the project (as are project inputs and outputs) and may undergo transformation at any stage. The optionality of the analysis is shown in Fig. 1 as a dotted line. The digital assets may provide raw data such as historical records as inputs into the project. The project artifacts may themselves be digital data that result from project work. Learning occurs through the project work as well as through investigating and understanding variances that may results in digital assets that are provided to project to produce project artifacts such as lessons learned documents or process improvements. Furthermore, the learnings may result in revisions to practices, models, or project artifacts.

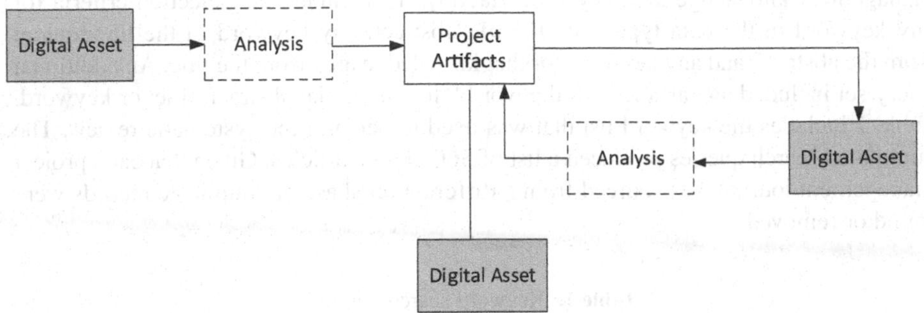

Fig. 1. Conceptual framework for the relationship between digital assets and project artifacts

The research methodology consisted of using a systematic review of the literature to compile the digital assets classification and an illustrative case to validate its relevance. "A systematic review is a review of a clearly formulated question that uses systematic and explicit methods to identify, select, and critically appraise relevant research, and to collect and analyze data from the studies that are included in the review" [27, p. 336]. A practitioner researcher-led case study was used to verify the relevance of the digital assets in projects.

3.2 Systematic Literature Review

Research Question. The research question for the systematic literature review is what digital assets are available through a project context? Thus, bodies of knowledge from the project management domain were selected to place boundaries around the research.

Keyword Selection. A literature review was performed to identify and classify digital assets in a project context. The project management bodies of knowledge that are used worldwide were reviewed to identify the project artifacts that could be digital assets: "ISO 21500:2012, Guidance on Project Management" [12], APM Body of Knowledge 6th Edition [11], A Guide to the Project Management Body of Knowledge (PMBOK guide) 6th Edition [10], and PRINCE2 [28]. Although criticized by some researchers, the "standards have come to represent an institutionalized collective identity of project managers" [29, p. 37]. Therefore, they offer guidelines for identifying project data sources. From the list of project artifacts, the keywords for the file content and knowledge areas were compiled into a list.

Article Selection. The main research journals that focus on project management [2] (i.e., Project Management Journal [PMJ], International Journal of Project Management

[IJPM], and IEEE Transactions on Engineering Management [IEEE] [2] and International Journal of Managing Projects in Business [IJMPB]) for the years 2000–2020 were selected for the keyword search. The bibliographic data from these journals were downloaded from the Emerald, ScienceDirect, IEEE Xplore, and Sage databases into the Endnote reference system. A set of search queries were created for each project management knowledge area keyword. Each query included the selection criteria for any keyword in the data type from the file abstract, any keyword in the file content from the abstract, and any keyword for that knowledge area from the title. An additional query set included any article with the word "digital" in the abstract, title, or keyword. Table 1 includes the keyword list that was used to perform the systematic review. The cumulated search queries produced a list of 360 unique articles. Given that each project management journal was sourced from a different database, no duplicate records were found or removed.

Table 1. Keyword search criteria

Search topic	Keywords
Data type	Image, text, numeric, static, dynamic, artificial, tool
File content	Register, report, data, calendar, file, list, system, log, estimate, forecast, brief
Procurement	Procurement, contract, legal, law, supplier, vendor, contractor
Financial	Financial, cost, budget, estimate, forecast, business case
Schedule	Calendar, schedule, milestone, duration, time, activities
Scope	Requirement, scope, assumptions, feature, user story, specification, acceptance
Resource	Resource, team, people, competence, competency
Risk	Risk, uncertainty
Integration	Change, configuration, control, life cycle, methodology, integrate
Communication	Communication, coordination
Stakeholder	Stakeholder, sponsor, owner, investor
Quality	Quality, test, evaluation, traceability

Article Screening. The abstracts for the selected articles were reviewed to determine if the article described the content or production of a digital project artifact. Based on the abstract review of the 360 articles, 99 articles were potentially relevant to the research topic. The full-text review produced 49 articles that described digital assets in sufficient detail to support the classification. To be included, the digital assets had to be a file or data item that could be 1) accessed by or created by a project actor or action, 2) be described in enough detail to determine its classification characteristics, and 3) be relevant to at least one of the project management knowledge areas. Based on observation, some items were included whenever a digital trace could be identified. For example, a case study that described an interview (which falls into the observation category) would normally be excluded; however, if a case study described a recording or transcription of an interview, the recording was included in the classification model. Table 2 provides an overview of the number of articles extract per journal and analysis stage. The classification details for the articles and digital assets are provided in the next section.

Coding Strategy. The coding strategy used to identify and classify digital assets was customized from the classification categories provided by [3, 26, 18]. In [3], digital assets were classified as digital communication between actors, virtual representations and models of project objects, or textual or numerical documents. That study created decision support monitoring processes based upon the physical attribute (e.g., size or dates), content or context (e.g., origin, project stage) of the digital asset. Those attributes were not considered in this study.

Quinton and Reynolds [26] specified the dimensions of the data, including data type (attitudinal or behavioral), distances from the data source (primary or secondary), data generation (mythically manufactured or naturally occurring), and data visibility (public or private). They also specified the characteristics of the dataset (big data, open data), the information (encoding format, provider), usage, and ethical challenges. In this study, we grouped digital assets using the data dimensions from [26] as well as attributes from our conceptual model.

Willems and Vanhoucke [18] provided a framework for describing the articles that added richness to their classification model on project control and earned value management. Their framework included the journal, research problem, contribution, methodology, analysis, and application of the paper. We analyzed the journal, methodology, data collection method, and analysis for each article used in the classification model.

After collecting and reviewing all articles within the defined scope, we compiled a list of digital assets that met the criteria and developed the classification framework for the articles and for the digital assets.

3.3 Illustrative Case Study

Case studies are a method of connecting qualitative understanding with issues, situations, and actions occurring in projects [30]. The case study method is consistent with the recommendation that practitioner-researchers familiar with scientific research and the professional practice should contribute knowledge to the project management discipline [31]. Furthermore, the case study approach provides multiple sources of evidence—archival records and participant observation [32]. This study included an illustrative case that involved extracting email and calendar data from Microsoft Outlook into Excel using a Visual Basic macro. The data were combined with the project contact list to identify roles and standardize the data. The data were anonymized in Excel. The Excel data were imported into R Studio version 1.3.1073 for analysis. Social network analysis, topic modeling, and reporting functions were used to analyze the data and present the results for the study. Detailed statistics are not reported in this study because the goal of the illustrative case was to demonstrate the use of digital assets from projects.

Table 2. Number of journal articles by stage

Journal	Total articles	Screened articles	Full-text review	With digital assets
IEEE	1401	124	35	17
IJMPB	644	54	14	5
IJPM	1918	139	38	19
PMJ	1080	43	12	8
Total	**4414**	**360**	**99**	**49**

3.4 Validity and Reliability

Validity refers to the extent to which the items are comprehensive and representative of the expected concept. Reliability estimates the degree to which the results are not random and free from unstable errors and reflect what is intended. The following steps were taken to ensure content external validity, internal consistency, relevance, and reliability. First, a conceptual model was designed through the theoretical lens of organizational learning and concepts from the triple-loop learning model [22]. The theories provided a framework for the use of data in defining learning goals and furthering project actions. Next, a systematic review was used to identify digital assets. The project management standards were used to identify keywords relevant to project management subjects. The keywords guided a systematic literature review of the dominant project management journals. Next, the coding strategies followed classification models for digital assets from research and project management [3, 18, 26]. The steps ensured the internal and external validity of the model. Finally, the research methodology used to define the classification model is documented in a manner that includes replication logic. The 27-item checklist and a four-phase flow from the PRISMA Statement were used to guide the study and report the results [27]. Care was taken to fully document the search process and the inclusion and exclusion criteria to support repeatability [33].

4 Research Findings

The articles from the systematic literature review were processed to identify digital assets available in a project context. Each digital asset was described and classified using the characteristics detailed in this section.

4.1 Classification Framework

The following characteristics were identified for each digital asset. The abbreviations presented here are used in the tables that describe the artifacts.

1. The *category* combines the digital assets into four groups: the communication between actors; virtual representation or models; and records and reports; and systems. The research found that the digital assets are embedded in computer systems

such as Computer-aided design (CAD), Geographical Information System (GIS), project management information system (PMIS), patent database, project scheduling, social media applications, virtual meeting platforms, or virtual reality technologies. The data that can be extracted as exports, database transactions, or tabular records from such systems are classified in the study [8]. Tables 3, 4 and 5 include the classification details of the assets by category.

2. The *digital asset* is a descriptive name for the data artifact.
3. *Data type* (DT) identifies the data as attitudinal or behavioral. Attitudinal data describes what people say, behavioral data describes what people do, or mixed.
4. The *data source* (DS) identifies the items as a primary source where raw data can be collected with a specific question in mind (e.g., an email) or a secondary source where the data has already been filtered or interpreted by someone like the project manager (e.g., a status) or a model.
5. *Visibility* (VIS) identifies the location and ownership of the data. The options include public, private, or open. Public data are accessible from a public location within the project environment; however, there may be access controls or restrictions; private data are confidential to a specific individual or group; open are public data from a public source such as local or government projects.
6. The *encoding* identifies the format of the data. The data can be text, numeric or relational data, images or spatial data, videos or audio, or a mixture.
7. The article *reference* (Ref), *analysis methods (Anal)*, and *project artifact* characteristics are given in the next sections.

4.2 Article Reference

There were 49 articles that described the input or outputs of project activities, digital project artifacts, or digital assets used in project research. The article reference is indicated for each of the digital assets. The articles approached the research population using statistical methods (Stat), case studies (CS), experiments (EXP), or literature reviews

Table 3. Digital assets—communications between actors

Asset	DT	DS	Vis	Encoding	Anal	Project artifact	Ref(s)
Contact lists	A	P	Private	Video, audio, text	Stat	COMM	[34]
Email - private	A	P	Private	Text	Stat	SE	[16]
Email - public	A	P	Public	Text	Obs	TE, TML	[35]
Media	A	P	Open	Video, audio	Txt	IL	[36]
Meeting reports	A	S	Private	Text	Txt	IL	[36]
Newspaper articles	A	P	Open	Text	Obs	FB	[37]
Parliament proposals	A	P	Open	Text	Obs	FB	[37]
Public comments	A	P	Open	Text	Comp	SR	[38]
Recordings	M	P	Private	Video, audio	Obs	TE	[39]
Social media groups	A	P	Private	Video, audio, text	Stat	COMM	[34]
Virtual meetings	A	P	Public	Video, audio, text	Txt	AL	[40]

Table 4. Digital assets—virtual representations

Asset	DT	DS	Vis	Encoding	Anal	Project artifact	Ref(s)
3-d or 4-d diagrams	B	P	Public	Image	Obs	TE	[39]
Aerial images	B	P	Public	Image	Comp	RR	[41]
Artificial intelligence models	B	S	Private	Numeric	AI	FC	[25, 42]
Building information modelling	B	P	Public	Relational data	Obs	PSCH	[43]
Competency matrix	B	S	Public	Text	Obs	RREQ	[44]
Earned value model	B	S	Private	Relational data	Obs	FC, EVMGT, RR, FS	[45]
Engineering change proposal	B	P	Public	Text	Comp	RR, CE, DE, PS	[46]
Mathematical models	B	S	Public	Numeric	Stat	RREQ	[47, 48]
Matrix	B	S	Private	Relational data	Comp	TS, AN	[49]
Network model	B	S	Public	Relational data	Stat	RR	[50]
Sensor data	B	P	Public	Relational data	Stat	AC	[51]
Spatial data	B	P	Public	Spatial data	Comp	PSCH	[52]
Time-based activity network	B	P	Public	Text	Stat	TAS	[53]
Virtual reality models	B	P	Private	Video, audio	Obs	TE	[39]

(LR). Statistical methods used a large sample from the population, case studies focused on understanding a few cases in detail, experiments were situations manufactured by the researcher, and structured reviews were topical studies from journal articles or other secondary sources. The data collection method defines the type of data that was used as a research or validation data source in the article; the definitions follow a model from [18]. Historical real-world data (Hist) are results from actual projects. Simulated-data (Sim) are generated from a simulation method such as a random sample. Qualitative data (Qual) represent an illustrative case or a single project case study.

Table 5. Digital assets—reports and records

Asset	DT	DS	Vis	Encoding	Anal	Project artifact	Ref(s)
Academic literature	A	S	Open	Text	AI	SSL	[54]
Budget	B	S	Private	Relational data	Obs	FB	[37]
Charts - graphs - drawings	B	S	Private	Image	Comp	PSCH	[52]
Chronological database	B	S	Public	Text	Obs	SR	[15]
Configuration commit	B	P	Public	Text	Stat	CI	[55]
Database transaction	B	P	Private	Relational data	Obs	FC, EVMGT, RR, FS	[45]
Fact sheets or annual reports	A	S	Open	Text	Stat	IP	[56]
Integrated master plan	B	S	Public	Text	Obs	PSCH	[57]
Item metadata	B	P	Public	Text	Stat	AN	[3]
Patent documents	A	P	Open	Text	Stat	PS, FC	[58]
PMIS-task assignment	B	P	Private	Relational data	Obs	TAS	[59]
PMIS-time-tracking	B	P	Private	Relational data	Obs	TAS	[59]
Project documents	M	S	Public	Text	Comp	QM	[42]
Resource allocation database	B	P	Private	Text	Obs	RC	[60]
Schedule and cost forecast transactions	B	S	Private	Relational data	Obs	FC, EVMGT, RR, FS	[45]
Searchable document database	M	P	Public	Text	Obs	CT	[14]
Survey responses	A	S	Public	Text	Stat	AC	[51, 61]
Tabular	B	S	Private	Relational data	Stat	RRPT	[61]
Technical reports	M	S	Private	Text	Comp	RR, CE, DE, PS	[46]

4.3 Analysis Methods

Type of analysis describes the techniques the article used for analyzing data in three broad categories: observational, computerized, and statistical. These definitions extend the model created by [18]. Observational analysis (Obs) includes data from literature surveys or empirical investigations; computerized analysis (Comp) includes automated systems or automated processes; statistical analysis (Stat) includes mathematical, network, or other analytical modeling methods; artificial intelligence (AI) includes methods that learn from experience and apply the learning to new situations; and text analysis (TXT) includes methods that analyze words or phrases from free-text.

4.4 Project Artifacts

The project artifacts that were inputs to or outputs from digital assets were documented in the classification model. Since the papers used a variety of names to describe similar content, artifact names from the project management standards were used whenever possible. In this section, the description of the artifacts is consolidated into a synthesis of the knowledge area from the project management standards [10–12, 28].

The integration knowledge area includes artifacts and processes that unify, consolidate, and integrate work across different areas [10, 12, 28]. A *change log* (CL) is a list of alterations submitted during the project and their status; it is used to communicate change requests to the impacted stakeholders. A *change request* (CR) is a proposal to modify a previous agreement for producing a work product, document, deliverable, or other work items. An *issue lists* (IL) is a catalog of unexpected events, problems, or tasks that are present in a project, and action must be taken to address the topic. The *lessons learned register* (LL) is the result of a project learning and may be produced as the result of a post-project or retrospective review.

The scope area includes artifacts to control what will and will not be included in the project work and may be defined by a business case [10, 28]. *Requirements* describe the needs of the customer and the *acceptance criteria* (AC) describe the performance indicators that will be used to judge success. The *product specifications* (PS) identify the customer's requirements and design parameters or attribute alternatives, including critical parameters such as materials, dimensions, product and process technology, testing, and the number of possible changes. The *configuration item* (CI) is a file that represents the basic unit of work in software engineering projects. *Work breakdown structure* (WBS) is a hierarchy-based list of all the work to be executed by the project team to create the necessary deliverables and meet the project objectives.

The financial and cost management artifacts include the monetary considerations needed to justify and control the project costs [11]. The forecasts include *financial forecasts* (FF) for the monetary costs at completion and include measures necessary to monitor the financial performance of the project. The *financial costs* (FC) identify where and when costs will be expended. The *earned value method* (EVM) is a typical method used to forecast and monitor project schedule and cost performance. The estimates include *cost estimates* (CE) for the total monetary costs for the project; *duration estimates* (DE) for the overall project, task, or change duration; and *effort estimates* (EE) for the amount of work.

Activity lists, activity networks, duration estimates, and schedules are artifacts identified in the schedule knowledge area, which focuses on the timely delivery and use of resources [10]. The *project schedule* (PSCH) is a date-based activity-level list that provides the basis for assigning resources and developing the budget. *Schedule forecasts* (FS) estimate the completion dates. The *activity list* (AL) is a breakdown of tasks or activities required to produce the project's work products or otherwise meet the project objectives.

The resource management knowledge area focuses on having the right people and material available [10, 12]. The *resource requirements* (RREQ) identify the labor, material, equipment, and supplies needed for the project activities and work products. The *resource calendar* (RC) identifies when each specific resource is available; it may include the availability by day of the week, hours per day, working times, shifts, public holidays, vacations, etc. The *team assignment* (TAS) identifies tasks and persons, groups, or organizations responsible for the tasks. The *team member list* (TML) identifies the people that are assigned to the project. The *team structure* (TS) depicts the hierarchy and relationship of the team members or the entire project organization.

The risk artifacts focus on managing the positive and negative uncertainty in the project [10–12, 28]. The *risk register* (RR) is a structured record, list, or document that details the identified uncertainties and their characteristics. The *risk report* (RRPT) provides *risk status* (RS) and summary information on the project risks and is used to communicate with the impacted stakeholders.

Project communications include the exchange of information between the project, its stakeholders, and team members [10, 12]. The communications are focused externally outside the project as *stakeholder engagement* (SE) or within the project as *team engagement* (TE). The *stakeholder register* (SR) contains details of the entities with interest in the project or its outcome, and it provides an overview of their roles and attitude toward the project.

The procurement knowledge area is centered around the contract and legal concerns and supplier management [10, 11]. A *contract* (CT) is an agreement reached between parties with the intention to create a legal relationship that protects the interests of the parties for consideration or understanding of an exchange of benefits and losses. *Intellectual property* (IP) are outcomes of the project that result in intangible products such as licenses, patents, brands, etc. The *selected supplier list* (SLR) includes preferred suppliers or contractors that were evaluated based upon established criteria and selected to provide products or services to the project.

Quality measures (QM) are performance metrics or measures that evaluate the performance of the project or the quality of the delivered product [10, 28].

4.5 Illustrative Cases

As an illustrative case for using the classification model, two questions were asked as a retrospective review of the case study project: Are we communicating enough with our stakeholders, and are we discussing the right topics? The research article from [62] were used to place boundaries around these two questions. The article identified factors frequency, content, and medium of communications between the project manager and the sponsor. The digital asset classification model identified email and meeting reports as being related to stakeholder engagement [16, 36] and the contact list and social media as related to communications. The researcher having access to project artifacts as the responsible project manager chose to use personal email and calendar transactions and the project contact list to address the questions. Meeting reports were distributed through email so were indirectly part of the analysis. Social media was not used during the project.

The subject, sender, receiver, and date were extracted from emails, and the subject, attendees, location, and date from calendar meetings were extracted from Outlook to an Excel file. The contact list for the project was used to assign each message sender and receiver and meeting participant with an organizational level and project role. Personal identification information was removed from further processing.

For the analysis of communication frequency, frequency counts were computed and coded as follows: daily, weekly, monthly, at milestone, at phase end, and ad hoc. For the analysis of the communication mediums, "written" was used for email, and "verbal" was used for calendar meetings. Face-to-face interactions were not considered since the project was conducted in 2020 (i.e., during the COVID-19 pandemic). For the analysis of the communication content, text mining was used on the email and calendar subjects. Specifically, topic modeling created categories of topics based on words in the email or calendar subject. For reporting in this study, the topics were manually identified in some categories from [62]: status, issues, next steps, work products and others.

Table 6 includes the selected communication patterns for written communications and Table 7 for verbal. The project was conducted between May and August-2020, lasting for approximately 13 weeks. It had seven phases and four milestones. There were 48 people in written communication with 8 different roles. The sponsor is one member of the steering committee (SteerCo) role. In verbal communication, 37 people were involved in 9 roles. The communication pattern suggests the project work product content was discussed with the sponsor when needed and through written communications, and weekly verbal meetings were used for status updates. While this communication pattern does not neatly fit with the frequency for content factors from [62], it does reveal that the sponsor was sufficiently engaged. In summary, the digital data assets used in this illustrative case answered key questions about the project execution. There are possibilities for analysis of communication patterns described by [9], which is currently beyond the scope of this case. Conversely, there are limitations created by the need to remove personally-identifying information to protect individual privacy.

Table 6. Selected written communication patterns

Role	Email communication pattern	Mail theme
Supplier	Adhoc + Some phase ends + Some milestones	Issues or open items
SteerCo	Adhoc + Some phase ends + Some milestones	Work product
Team-functional	Adhoc + Some phase ends + Some milestones	Work product

Table 7. Selected verbal communication patterns

Role	Calendar communication pattern	Main theme
Supplier	Weekly + Some phase ends + Some milestones	Next steps
SteerCo	Weekly + All phase ends + Some milestones	Status
Team-functional	Weekly + Some phase ends + All milestones	Next steps

5 Discussion

This collection of data assets identified in this study underscores the variety of content, and consequently, management challenges that are inherent in projects. It validates some of the proposed use cases for digital assets and uncovers where value may be found in the digital assets.

The 11 items for communication category between actors represent textual documents or video or audio recordings that capture the attitudes of individuals and organizations. In most cases, they represent a primary source of data that can be analyzed using a multitude of analysis methods such as network analysis, statistics, or text mining. The strong showing of assets in the communication category support the argument made by [8] that communication software is important in project realization environments. Conversely, the variety of sources suggest that software is only one tool for understanding communication patterns. Additional assets such as email, meeting reports, or public documents should be used in the analysis of communication frequencies or analysis of any of the 11 project communication patterns identified by [9]. However, one of the challenges in accessing and analyzing communication data is preserving privacy and compliance with privacy laws [26, 51, 63].

The reports and records category of 19 items is a mixed bag where it is more likely that the acts and actions within a project can be analyzed. Many of the assets could be extracted from project management software or databases. However, digital traces of actual behaviors are also available in open source or other non-traditional locations. For example, [53] analyze configuration commit metadata of an engineering design process data of a biomass power plant to determine team assignment.

The virtual representation and model's category introduced 14 data assets that all focus on behavior. This is a category of non-traditional project management data sources such as 3D and 4D models, spatial data, and other virtual models. The analytic methods in this category tend toward virtual and visual methods. For example, [41] used drones to capture aerial images of a small tank farm in Termini Imerese, Italy, to create a risk register. Cheng and Roy [25] and [54] used simulated or linguistic data to create artificial intelligence models that are characterized in the model as digital assets.

The final group is computer systems, databases, and applications that generate the data. Of the 8 assets identified, three are traditional project management systems, according to the details from [8]. However, building information systems that use spatial data and 3D and 4D diagrams in preparing project schedules are a modern development in construction project management [52].

From the project management knowledge area perspective, the quality, scope, and stakeholder areas were represented by the fewest number of assets and communication, schedule, and integration by the most. However, the financial and integration areas were the most dynamic with artifacts in all four categories. Example uses of assets for integration include the experimental use of virtual reality and recordings as digital boundary objects [39] or innovation cases where 3D and 4D diagrams are used to define team structures and activity networks [64]. Both the stakeholder and the risk knowledge areas used project and other documents for building stakeholder and risk registers [15, 65].

The articles used to compile the asset list offer examples of how data is used to analyze behaviors and attitudes. Furthermore, the existence of the assets may provide an input into the activity monitoring. Snider, Gopsill, Jones, Emanuel, and Hicks [3] described four monitoring use cases for digital assets, including monitoring communication trends coordinated with the project schedule, predicting time-to-completion, and identifying system and module dependencies. The case illustrated in this study demonstrate that the everyday project activities of scheduling meetings and sending emails provide an operationalized view of trends and gaps in communication. This confirms that project performance can be evaluated from an analysis of the data.

There are two benefits the digital assets offer for academic research and to project lessons learned. They make the knowledge objective and independent of individual memories or recalls. This removes human cognition and social interactions from the data collection process, which moves from self-reports to objective observation. Second, it avoids bias in the data due to individuals moderating their behavior due to observations. It does, however, introduce other issues such as gaining access to the data and overcoming privacy and ethical concerns [26, 51, 63]. Nevertheless, the digital assets provide the benefits of direct measurement that is less intrusive than other approaches such as surveys [24, 26].

The identification of assets, such as the use of spatial data and 3D and 4D models, highlights that the data generated are not just one-dimensional that have value during the project but are also items that can provide value after the project is completed. The assets are also not only inwardly focused. For example, [58] used the patent database to define a product specification in the thin film transistor-liquid crystal display market that would be financially lucrative. This can explain why [14] determined that in new digital project models, project teams are contractually obliged to deliver digital information to the long-term owner or operator. Similarly, [24] detailed the competitive advantages derived from using computer systems for knowledge sharing between consumers and organizations; similar, benefits would apply to knowledge exchange between stakeholders and projects.

The study identified the convergence of digital information and project practices as a future area of research.

6 Conclusion

Projects involve a variety of tools, methods, and systems for their efficient management. Consequently, they produce valuable digital data and individual and organizational learnings. The research identified four categories and 52 items of digital assets that are available throughout projects. The assets provide a source for primary and secondary data and can be used for project studies or for creating data-driven project management processes. Furthermore, they are useful for research and integrating individual learnings into organizations, accessing project learning, and providing a foundation for project management decision-making and artificial intelligence systems. The illustrative case provided an example for using readily available digital information to provide transparency and learnings from project actions.

6.1 Contribution to Knowledge

By consolidating the list of digital assets, this study contributes to the project management literature on communications, lessons learned, and digitalization. This digital asset classification fills a gap in the literature by providing a comprehensive list of digital assets available from a variety of project contexts. This work extends the examples of digital assets provided by [3]. There has been a call for an improvement in project management processes and methods for disseminating organizations' learnings. The model should help reveal new possibilities for analysis to make project management lessons learned more effective, complete, and structured [13, 22]. Moreover, it provides a platform for moving knowledge from the subjective to the objective and reducing the dependence on only individual learnings [22, 23]. Finally, this research fits into the trend of using artificial intelligence and traditional statistical methods to solve project management problems [3, 18, 25]. Specifically, artificial intelligence is usually realized through the development of algorithms and models that require training data. The digital assets identified herein may be the sources of that data.

6.2 Implication for Practice

The practical implications are a list of digital assets that can be inputs in the creation of project artifacts and sources for monitoring and controlling project activities and for sense-making in retrospectives or lessons learned. Moreover, this categorization is useful for decision support and artificial intelligence systems model development that requires raw data. For example, [17] provided models for using social networks to improve project management performance and the illustrative example provided a method to extract social network data from project communication artifacts. Next, collecting lessons learned and the analysis and synthesis of team behaviors and attitudes are also possible by using the complication of digital assets. This should support implementing improvement processes as promoted by the project management standards [10–12, 28]. Finally, the digital assets themselves are valuable work products that can provide contractual value to project clients and sponsors, as argued by [14]. However, the project size, team competency, or project manager capabilities will be factors that determine the availability of data, and the need and ability to use digital assets in a project situations or research studies. Therefore, the digital assets proposed by the study may be too sophisticated for routine usage.

6.3 Implication for Research

Projects offer a rich environment for the time and actors, which are usually fixed at the start of the project. Thus, they are ripe for applying multiple research methods such as action research, case studies, and experiments. Digital assets support tracing individual, group, and organizational behaviors. Furthermore, digital data are especially relevant as organizations transition to digital and remote working environments. This categorization offers academic researchers a catalog of data sources and analysis methods for studying complex project phenomena. However, there may be some challenges in gaining permission and clearance to utilize the data in the desired method. In addition, ethical use is a concern when dealing with data related to individuals [26].

6.4 Limitations and Further Research

This research was based on a systematic literature review at a single point in time and focused on a small selection of publications. The journal selection was limited to the main project management research journals identified by [2]; however, other data sources may have produced a wider or different selection of digital assets. The contribution was made by an individual researcher; this approach may have limited the identification of digital assets based on the knowledge or biases of the researcher. Furthermore, the research did not consider project size. Thus, the data assets may not be available in meaningful volumes or readily accessible in small projects.

Further research with project and organizational actors is needed to expand on the types of digital assets and further classify the data. An interesting extension would be to add the attributes relevant to each digital asset to the classification model. Moreover, determining a strategy for widening the selection of journal used and still providing a meaningful context would be interesting. Alternatively, other qualitative methods such as case studies could also be considered for the research. Finally, the rapid move to remote working driven by the COVID-19 pandemic offers an interesting avenue to research the impact of digital communications on projects.

References

1. Lundin, R.A., Söderholm, A.: A theory of temporary organization. Scand. J. Manage. **11**, 437–455 (1995). https://doi.org/10.1016/0956-5221(95)00036-U
2. Drouin, N., Müller, R., Sankaran, S.: Novel Approaches to Organizational Project Management Research: Translational and Transformational. Copenhagen Business School Press, Denmark (2013)
3. Snider, C., Gopsill, J.A., Jones, S.L., Emanuel, L., Hicks, B.J.: Engineering project health management: a computational approach for project management support through analytics of digital engineering activity. IEEE Trans. Eng. Manage. **66**(3), 325–336 (2019). https://doi.org/10.1109/TEM.2018.2846400
4. Jensen, A., Thuesen, C., Geraldi, J.: The projectification of everything: projects as a human condition. Proj. Manage. J. **47**(3), 21–34 (2016). https://doi.org/10.1177/875697281604700303
5. Lundin, R.A.: Project society: paths and challenges. Proj. Manage. J. **47**(4), 7–15 (2016). https://doi.org/10.1177/875697281604700402

6. Geraldi, J., Söderlund, J.: Project studies and engaged scholarship: directions towards contextualized and reflexive research on projects. Int. J. Manag. Proj. Bus. **9**(4), 767–797 (2016). https://doi.org/10.1108/IJMPB-02-2016-0016
7. Besner, C., Hobbs, B.: An empirical identification of project management toolsets and a comparison among project types. Proj. Manage. J. **43**(5), 24–46 (2012). https://doi.org/10.1002/pmj.21292
8. Muszyńska, K., Swacha, J.: Selecting project communication management software using the weighted regularized hasse method. In: Ziemba, E. (ed.) Information Technology for Management. Ongoing Research and Development. ISM 2017, AITM 2017. LNBIP, vol. 311, pp. 212–228. Springer, Cham (2018). https://doi.org/10.1007/978-3-319-77721-4_12
9. Muszyńska, K.: Project communication management patterns. In: 2016 Federated Conference on Computer Science and Information Systems (FedCSIS), pp. 1179–1188 (2016)
10. PMI: A Guide to the Project Management Body of Knowledge (PMBOK Guide) –Sixth Edition. Project Management Institute, Inc., Newtown Square, Pennsylvania, United States (2017)
11. APM: APM Body of Knowledge Sixth Edition. Association for Project Management, Buckinghamshire, United Kingdom (2012)
12. ISO: ISO 21500: 2012 Guidance on project management. International Standards Organization, Geneva, Switzerland (2012)
13. Duffield, S., Whitty, S.J.: Developing a systemic lessons learned knowledge model for organisational learning through projects. Int. J. Proj. Manage. **33**(2), 311–324 (2015). https://doi.org/10.1016/j.ijproman.2014.07.004
14. Whyte, J.: How digital information transforms project delivery models. Proj. Manage. J. **50**(2), 177–194 (2019). https://doi.org/10.1177/8756972818823304
15. Missonier, S., Loufrani-Fedida, S.: Stakeholder analysis and engagement in projects: from stakeholder relational perspective to stakeholder relational ontology. Int. J. Proj. Manage. **32**(7), 1108–1122 (2014). https://doi.org/10.1016/j.ijproman.2014.02.010
16. Hossain, L., Wu, A.: Communications network centrality correlates to organisational coordination. Int. J. Proj. Manage. **27**(8), 795–811 (2009). https://doi.org/10.1016/j.ijproman.2009.02.003
17. Toomey, L.: Social networks and project management performance: how do social networks contribute to project management performance? In: Paper presented at PMI® Research and Education Conference, Project Management Institute, Limerick, Munster, Ireland (2012)
18. Willems, L.L., Vanhoucke, M.: Classification of articles and journals on project control and earned value management. Int. J. Proj. Manage. **33**(7), 1610–1634 (2015). https://doi.org/10.1016/j.ijproman.2015.06.003
19. Nemati, H.R., Todd, D.W., Brown, P.D.: A hybrid intelligent system to facilitate information system project management activities. Proj. Manage. J. **33**(3), 42–52 (2002). https://doi.org/10.1177/875697280203300306
20. Jugdev, K., Perkins, D., Fortune, J., White, D., Walker, D.: An exploratory study of project success with tools, software and methods. Int. J. Manag. Proj. Bus. **6**(3), 534–551 (2013). https://doi.org/10.1108/IJMPB-08-2012-0051
21. Hazır, Ö.: A review of analytical models, approaches and decision support tools in project monitoring and control. Int. J. Proj. Manage. **33**(4), 808–815 (2015). https://doi.org/10.1016/j.ijproman.2014.09.005
22. McClory, S., Read, M., Labib, A.: Conceptualising the lessons-learned process in project management: towards a triple-loop learning framework. Int. J. Proj. Manage. **35**(7), 1322–1335 (2017). https://doi.org/10.1016/j.ijproman.2017.05.006
23. Scarbrough, H., Bresnen, M., Edelman, L.F., Laurent, S., et al.: The processes of project-based learning: an exploratory study. Manage. Learn. **35**(4), 491–506 (2004). https://doi.org/10.1177/1350507604048275

24. Ziemba, E., Eisenbardt, M., Mullins, R.: Use of information and communication technologies knowledge sharing by Polish and UK-based Prosumers. In: Ziemba, E. (ed.) Information Technology for Management: New Ideas and Real Solutions. LNBIP, vol. 277, pp. 49–73. Springer, Cham (2017). https://doi.org/10.1007/978-3-319-53076-5_4
25. Cheng, M.-Y., Roy, A.F.V.: Evolutionary fuzzy decision model for cash flow prediction using time-dependent support vector machines. Int. J. Proj. Manage. **29**(1), 56–65 (2011). https://doi.org/10.1016/j.ijproman.2010.01.004
26. Quinton, S., Reynolds, N.: Understanding research in the digital age. Sage, London (2018)
27. Moher, D., Liberati, A., Tetzlaff, J., Altman, D.G.: Preferred reporting items for systematic reviews and meta-analyses: the PRISMA statement. Int. J. Surg. **8**(5), 336–341 (2010). https://doi.org/10.1016/j.ijsu.2010.02.007
28. Siegelaub, J.M.: How PRINCE2 can complement PMBOK and your PMP. In: PMI® Global Congress 2004—North America. Project Management Institute, Anaheim, CA (2004)
29. Eskerod, P., Huemann, M.: Sustainable development and project stakeholder management: what standards say. Int. J. Manag. Proj. Bus. **6**(1), 36–50 (2013). https://doi.org/10.1108/17538371311291017
30. Floricel, S., Banik, M., Piperca, S.: The triple helix of project management research: Theory development, qualitative understanding and quantitative corroboration. In: Drouin, N., Müller, R., Sankaran, S. (eds.) Novel Approaches to Organizational Project Management Research: Translation and Transformational, pp. 402–429. Copenhagen Business School Press, Norway (2013)
31. Lalonde, P., Bourgault, M.: Project, project theories, and project research: a new understanding of theory, practice and education for project management. In: Drouin, N., Müller, R., Sankaran, S. (eds.) Novel Approaches to Organizational Project Management Research: Translation and Transformational, pp. 430–451. Copenhagen Business School Press, Norway (2013)
32. Yin, R.K.: Case Study Research: Design and Methods, 5th edn. SAGE Publications Inc, Thousand Oaks (2014)
33. Kitchenham, B., Brereton, P., Li, Z., Budgen, D., Burn, A.: Repeatability of systematic literature reviews. In: 15th Annual Conference on Evaluation & Assessment in Software Engineering (EASE 2011), pp. 46–55 (2011). https://doi.org/10.1049/ic.2011.0006
34. Zhang, Y., Sun, J., Yang, Z., Wang, Y.: Mobile social media in inter-organizational projects: Aligning tool, task and team for virtual collaboration effectiveness. Int. J. Proj. Manage. **36**(8), 1096–1108 (2018). https://doi.org/10.1016/j.ijproman.2018.09.003
35. Lee-Kelley, L., Sankey, T.: Global virtual teams for value creation and project success: a case study. Int. J. Proj. Manage. **26**(1), 51–62 (2008). https://doi.org/10.1016/j.ijproman.2007.08.010
36. Gil, N.A.: Language as a resource in project management: a case study and a conceptual framework. IEEE Trans. Eng. Manage. **57**(3), 450–462 (2010). https://doi.org/10.1109/TEM.2009.2028327
37. Knardal, P.S., Pettersen, I.J.: Creativity and management control – the diversity of festival budgets. Int. J. Manag. Proj. Bus. **8**(4), 679–695 (2015). https://doi.org/10.1108/IJMPB-11-2014-0082
38. Sperry, R.C., Jetter, A.J.: A systems approach to project stakeholder management: fuzzy cognitive map modeling. Proj. Manage. J. **50**(6), 699–715 (2019). https://doi.org/10.1177/8756972819847870
39. Alin, P., Iorio, J., Taylor, J.E.: Digital boundary objects as negotiation facilitators: spanning boundaries in virtual engineering project networks. Proj. Manage. J. **44**(3), 48–63 (2013). https://doi.org/10.1002/pmj.21339
40. Iorio, J., Taylor, J.E.: Precursors to engaged leaders in virtual project teams. Int. J. Proj. Manage. **33**(2), 395–405 (2015). https://doi.org/10.1016/j.ijproman.2014.06.007

41. Aiello, G., Hopps, F., Santisi, D., Venticinque, M.: The employment of unmanned aerial vehicles for analyzing and mitigating disaster risks in industrial sites. IEEE Trans. Eng. Manage. **67**, 1–12 (2020). https://doi.org/10.1109/TEM.2019.2949479
42. Duman, G.M., El-Sayed, A., Kongar, E., Gupta, S.M.: An intelligent multiattribute group decision-making approach with preference elicitation for performance evaluation. IEEE Trans. Eng. Manage. **67**(3), 1–17 (2020). https://doi.org/10.1109/TEM.2019.2900936
43. Papadopoulos, T., Ojiako, U., Chipulu, M., Lee, K.: The criticality of risk factors in customer relationship management projects. Proj. Manage. J. **43**(1), 65–76 (2012). https://doi.org/10.1002/pmj.20285
44. Marnewick, C., Marnewick, A.L.: The demands of industry 4.0 on project teams. IEEE Trans. Eng. Manage. **67**(3), 1–9 (2020). http://dx.doi.org/10.1109/TEM.2019.2899350
45. Batselier, J., Vanhoucke, M.: Construction and evaluation framework for a real-life project database. Int. J. Proj. Manage. **33**(3), 697–710 (2015). https://doi.org/10.1016/j.ijproman.2014.09.004
46. Yeasin, F.N., Grenn, M., Roberts, B.: A Bayesian networks approach to estimate engineering change propagation risk and duration. IEEE Trans. Eng. Manage. **67**(3), 1–16 (2020). https://doi.org/10.1109/TEM.2018.2884242
47. Ghomi, S.M.T.F., Ashjari, B.: A simulation model for multi-project resource allocation. Int. J. Proj. Manage. **20**(2), 127–130 (2002). https://doi.org/10.1016/S0263-7863(00)00038-7
48. Fathian, M., Saei-Shahi, M., Makui, A.: A new optimization model for reliable team formation problem considering experts' collaboration network. IEEE Trans. Eng. Manage. **64**(4), 586–593 (2017). https://doi.org/10.1109/TEM.2017.2715825
49. Browning, T.R.: Applying the design structure matrix to system decomposition and integration problems: a review and new directions. IEEE Trans. Eng. Manage. **48**(3), 292–306 (2001). https://doi.org/10.1109/17.946528
50. Hwang, W., Hsiao, B., Chen, H.-G., Chern, C.-C.: Multiphase assessment of project risk interdependencies: evidence from a University ISD project in Taiwan. Proj. Manage. J. **47**(1), 59–75 (2016). https://doi.org/10.1002/pmj.21563
51. Hossain, M.M., Prybutok, V.R.: Consumer acceptance of RFID technology: an exploratory study. IEEE Trans. Eng. Manage. **55**(2), 316–328 (2008). https://doi.org/10.1109/TEM.2008.919728
52. Bansal, V.K., Pal, M.: Construction schedule review in GIS with a navigable 3D animation of project activities. Int. J. Proj. Manage. **27**(5), 532–542 (2009). https://doi.org/10.1016/j.ijproman.2008.07.004
53. Parraguez, P., Piccolo, S.A., Perišić, M.M., Štorga, M., Maier, A.M.: Process modularity over time: modeling process execution as an evolving activity network. IEEE Trans. Eng. Manage. **PP**, 1–13 (2019). https://doi.org/10.1109/TEM.2019.2935932
54. Movahedian Attar, A., Khanzadi, M., Dabirian, S., Kalhor, E.: Forecasting contractor's deviation from the client objectives in prequalification model using support vector regression. Int. J. Proj. Manage. **31**(6), 924–936 (2013). https://doi.org/10.1016/j.ijproman.2012.11.002
55. Nan, N., Kumar, S.: Joint effect of team structure and software architecture in open source software development. IEEE Trans. Eng. Manage. **60**(3), 592–603 (2013). https://doi.org/10.1109/TEM.2012.2232930
56. Michelino, F., Lamberti, E., Cammarano, A., Caputo, M.: Open innovation in the pharmaceutical industry: an empirical analysis on context features, internal R&D, and financial performances. IEEE Trans. Eng. Manage. **62**(3), 421–435 (2015). https://doi.org/10.1109/TEM.2015.2437076
57. Chang, A., Hatcher, C., Kim, J.: Temporal boundary objects in megaprojects: mapping the system with the integrated master schedule. Int. J. Proj. Manage. **31**(3), 323–332 (2013). https://doi.org/10.1016/j.ijproman.2012.08.007

58. Yoon, B., Park, Y.: Development of new technology forecasting algorithm: hybrid approach for morphology analysis and conjoint analysis of patent information. IEEE Trans. Eng. Manage. **54**(3), 588–599 (2007). https://doi.org/10.1109/TEM.2007.900796
59. Braglia, M., Frosolini, M.: An integrated approach to implement project management information systems within the extended enterprise. Int. J. Proj. Manage. **32**(1), 18–29 (2014). https://doi.org/10.1016/j.ijproman.2012.12.003
60. Celkevicius, R., Russo, R.F.S.M.: An integrated model for allocation and leveling of human resources in IT projects. Int. J. Manag. Proj. Bus. **11**(2), 234–256 (2018). https://doi.org/10.1108/IJMPB-09-2016-0074
61. Crispim, J., Silva, L.H., Rego, N.: Project risk management practices: the organizational maturity influence. Int. J. Manag. Proj. Bus. **13**(1), 187–210 (2019). https://doi.org/10.1108/IJMPB-10-2017-0122
62. Müller, R.: Determinants for external communications of IT project managers. Int. J. Proj. Manage. **21**(5), 345–354 (2003). https://doi.org/10.1016/S0263-7863(02)00053-4
63. Schwarzbach, B., Glöckner, M., Schier, A., Robak, M., Franczyk, B.: User specific privacy policies for collaborative BPaaS on the example of logistics. In: 2016 Federated Conference on Computer Science and Information Systems (FedCSIS), pp. 1205–1213 (2016)
64. Taylor, J.E., Levitt, R.: Innovation alignment and project network dynamics: an integrative model for change. Proj. Manage. J. **38**(3), 22–35 (2007). https://doi.org/10.1002/pmj.20003
65. Lehtiranta, L.: Risk perceptions and approaches in multi-organizations: a research review 2000–2012. Int. J. Proj. Manage. **32**(4), 640–653 (2014). https://doi.org/10.1016/j.ijproman.2013.09.002

Data Quality Model-Based Testing of Information Systems: Two-Level Testing of the Insurance System

Anastasija Nikiforova(✉), Janis Bicevskis, Zane Bicevska, and Ivo Oditis

Faculty of Computing, University of Latvia, Riga, Latvia
{anastasija.nikiforova,janis.bicevskis,zane.bicevska, ivo.oditis}@lu.lv

Abstract. In order to develop reliable software, its operating must be verified for all possible cases of use. This can be achieved, at least partly, by means of a model-based testing (MBT), by establishing tests that check all conditions covered by the model. This paper presents a Data Quality Model-based Testing (DQMBT) using the data quality model (DQ-model) as a testing model. The DQ-model contains definitions and conditions for data objects to consider the data object as correct. The proposed testing approach allows complete testing of the conformity of the data to be entered and the data already stored in the database. The data to be entered shall be verified by means of predefined pre-conditions, while post-conditions verify the allocation of the data into the database. The paper demonstrates the application of the proposed solution to the insurance system, concluding that it is able to identify previously undetected defects even after years of operating the IS. Therefore, the proposed solution can be considered as an effective complementary testing approach capable to improve the quality of an information system significantly. In the context of this study, we also address the MBT approach and the main factors affecting its popularity and identify the most popular ways of classifying MBT approaches.

Keywords: Complete test set · Data quality model · Information system · Model-Based testing · Pre-condition · Post-condition

1 Introduction

The main goal of developers over the years is to develop reliable software that can be used in a real life. Software testing aimed at contributing to this goal has attracted researchers and practitioners since software development began. It became an essential part of software development that requires significant resources. But, despite that, this challenge has not been resolved yet.

Model-based testing (MBT) is one of the most popular testing approaches in the practice and literature [1], following by such testing techniques as search-based, mutation, symbolic execution, combinatorial etc. In recent years, a high number of MBT approaches was presented [2–6], but most of them are presented without an execution

E. Ziemba and W. Chmielarz (Eds.): ISM 2020/FedCSIS-IST 2020, LNBIP 413, pp. 25–44, 2021.
https://doi.org/10.1007/978-3-030-71846-6_2

context or in an artificial environment, making it difficult to understand if they can be used for testing real systems. But probably the most disputable question is the choice of a model that highly affects the effectiveness of testing.

The aim of the study is to propose an alternative MBT approach that uses the data quality model (DQ-model) as a test model, called data quality model-based testing approach - DQMBT. As one of the primary tasks of information system (IS) is to collect and process data [7, 8], the data objects to be entered must be verified on their correctness described by conditions of data objects values. The correct data objects can be stored in a database (DB), while the information about the incorrect data objects must be provided to the data owner for data analysis, editing and re-entering them. Testing with these test cases therefore checks the accuracy of entering and storing data in both syntactic and contextual terms. The verification of the data to be entered is carried out by means of pre-conditions, whilst the data deployment in the DB is addressed through post-conditions. Conditions on values for the attributes of data objects are proposed to be used to prepare test cases that will process all correct and incorrect cases. If IS works correctly on tests that cover all elements of the model, it is assumed that complete system testing has been performed according to the selected model. The study therefore proposes a new complete testing criterion - verifying the correctness of all input data and its allocation in the DB with tests that check all possible input values conditions. The checks described in the DQ-model are not related to an implementation in the particular environment. The use of DQ-model aims to solve at least partly the limitations of the UML and complexity characterized to OCL [7], as well as suppose by-default covering both correct and incorrect cases in contrast with other approaches such as [9]. Using the DQ-model as a test model allows preparing test cases constructively for the verification of all conditions. This means that the proposed approach covers only part of the functional testing of IS, more precisely complete testing of the input data and the verification of the relevance of data stored in the DB with input data, but this would significantly improve the overall quality of IS, which are today one of the most important challenges [10, 11].

This paper is an extended version of the paper originally presented in the Federated Conference on Computer Science and Information Systems [12]. While the original paper addressed mainly the set of basic concepts that are used in the DQMBT approach, providing a "toy" example, this paper provides an in-depth study on the concepts, which were not thoroughly covered in the previous paper, involves the concepts of pre- and post-conditions, focusing on the application of the proposed approach to the real system of insurance policies. This allows testing the proposed approach on a real information system, leading to an initial set of quantitative and qualitative results.

Therefore, the following research questions were established: *(RQ1) What are a model-based testing success factors and how are the existing approaches classified? (RQ2) Is the previously proposed data object-oriented data quality model suitable for model-based testing? (RQ3) Is the proposed data quality model-based testing suitable for real system testing?* This question should be answered by applying DQMBT to a real system, using the test result, i.e. whether defects will be revealed in a selected component that had already been in industrial operation for a long period.

The paper is structured as follows: Sect. 2 forms a theoretical background addressing MBT, its popularity and classifications, Sect. 3 describes the methodology, presenting the algorithm and DQ-model, Sect. 4 presents analysis of the proposed solution applying it to the real system, Sect. 5 provides the study's contribution, implications and limitations, as well as conclusions in Sect. 6.

2 Theoretical Background

This section deals with the concept of MBT aimed to answer the RQ1 - *What are a model-based testing success factors and how are existing approaches classified?*

2.1 Basics of Model-Based Testing

Let us briefly address the concept of model-based testing (MBT) and the essential aspects to be taken into account proposing its derivatives. As already mentioned, the MBT is one of the most popular testing approaches in the practice and literature [1]. This trend can be easily explained by the list of advantages of MBT that varies in terms of its length and nature. However, perhaps the most commonly mentioned benefits are: (a) reduction of the test effort, especially the effort for test maintenance; (b) improved quality of the specification artifacts; (c) higher number of defects found at an early stage of the project due to model creation; (d) improved test case quality due to systematic coverage of the model and removal of redundancies and (e) improved quality of the system under test (SUT) due to faster and more efficient testing [13, 14]. Another important advantage of MBT is the time spent for testing. According to [14], the total number of hours spent on MBT testing is more than 3 times less than the manual testing, more than twice as low as the replay, and 1.9 and 1.17 times lower than scripting and keyword testing. In addition, although MBT is primarily aimed at functional testing, it can also be used to test some kinds of robustness such as testing with invalid or unexpected inputs that is the case for this study, unavailability of dependent applications, hardware or network failures [15], and even performance testing. It is another reason of its popularity – its universality.

Though besides the benefits, there are some limitations, too. First, the establishing of a MBT is quite resource consuming, so it is not advantageous if the number of test cycles is only one or two. However, the higher is the number, the higher is the appropriateness of MBT for testing purposes; the case described in this paper proves that. Second, quality and the level of detail of the specification artifacts and the type of the model selected and lying under the MBT process are vital, given that MBT is likely the combination of software modeling and testing at all levels (in line with [16]). In order to deal with this challenge, the previously proposed data quality model [17] is used to model the part of the system to be tested in this study.

Let us briefly address the key concepts of MBT. The MBT is traditionally characterized by 5 interrelated steps which may vary slightly depending on the study. We provide two the most common 5-step sets summarized in Table 1. While the first set appears to be more popular (2nd column in Table 1) and is well-accepted in many studies, such as [14, 15, 18–20], the 3rd column corresponds to the steps defined by Jorgensen [16] who is one of the leading researchers in the software testing and the MBT in particular. It is important that mainly the 2nd and 3rd steps differ, while steps 4 and 5 (executing tests and analysis of results) are common for the most traditional testing techniques, but the

1st step referred to the modeling the system is the core of the MBT. In this study we are following the most traditional set provided in the 2nd column.

Table 1. 5 steps of MBT.

#	5 steps of MBT [14, 15, 18–20]	5 steps of MBT by [16]
1	Model the system and/ or its environment	Model the system
2	Generate abstract tests from the model	Identify threads of system behavior
3	Concretize the abstract tests to make them executable	Transform these threads into test cases
4	Execute the tests on the SUT and assign verdict	Execute the test cases on the SUT and record the results
5	Analyze the test result	Revise the model and repeat the process

The central concept in MBT is a model because its adequacy depends on the accuracy of the SUT model. The best option is to give preference to the formal modeling notation, which has precise semantics (in line with [15]). As in the case of MDA, the UML class diagram can be considered to be one of the most common options for such a task, which is often considered to be a standard [2–4, 8, 14, 15]. However, according to [21], while this is a common assumption, the state of the art shows that UML is used in less than 30% cases. Perhaps this is related to the fact that neither UML nor the informal use case diagram are sufficiently precise or detailed for MBT, because it requires a description of the system's dynamic behavior that cannot be achieved using the traditional UML without extensions. There is a number of options for obtaining an extended UML model suitable for MBT that can be achieved by adding detailed information on the behavior of the methods in the class diagram; two of them are OCL (Object Constraint Language) post conditions [22] and state machine diagrams [23].

The indisputable advantage of the OCL is the ability to define constraints or specify pre- or post-conditions for operations. However, this makes it considerably more complicated, and every time there is a need for change, a qualified person should be involved. We believe that we can provide modeling capabilities, including pre- and post-conditions, without significant increase of complexity of our model. However, this has been achieved through its simplification and adopting it to the task assigned.

One of the most common solutions is to look for other technique, considering (1) a small size of the model and (2) its level of detail that can sometimes be conflicting. Therefore it is important to carry out engineering task to (a) decide *which system characteristics should be modeled to achieve the test objectives*, (b) *how much detail is useful*, and (c) *which modeling notation can express these characteristics most naturally* [8]. In the light of all the above, we decided to use the data object-driven data quality model (DQ-model) previously proposed in [17], which is based on the use of DSLs by adapting it to the given study. The use of DSL seems appropriate for the given study, which is also considered to be one of the most popular choices for MBT. [5] shows that DSL is the second most popular choice for MBT.

2.2 A Brief Overview on the Existing MBT Approaches

This paper does not suppose an in-depth analysis of the existing MBT approaches, considering that this was the main focus for recent studies such as [2–6], where 72, 22, 14, 36 and 87 papers respectively where thoroughly addressed by classifying MBT approaches by different dimensions, highlighting their pros and cons. Its popularity is also reflected in [21] and [24], where more than 200 MBT approaches were analyzed, dividing them by (a) the modeling language used and (b) the perspective from which the model represents information. Let us briefly address them.

[2] mainly addresses search-based techniques for model-based testing classifying MBT approaches by such aspects as (a) purpose - regression, structural, functional, GUI or stress testing, (b) evaluation – baseline, effective measure, cost measures, dataset, statistical test, (c) solutions such as model transformation, optimization process, landscape visualizations, type of search-based techniques, constraint handling, fitness function and (d) problems, i.e. application domain, model, test level, quality attributes, constraints, dimensionality, adequacy criteria.

Iqbal et al. [3] provides an overview on 87 papers focusing on the quality of delivery of the proposed MBT approaches. They conclude that it is generally low, mainly trying to comply with the guidelines for papers in scope of their sections, and less attention is paid to solution-specific details. In addition, most papers dedicated to MBT are presented without an execution context or in an artificial environment. In contrast, our proposal belongs to one of the most unique categories – industrial execution, i.e. in the real environment. Gurbuz et al. [5], however, highlight that most of the articles discuss MBT for particular application domains, such as automotive and railway, which seems to impact the MBT process.

Villalobos et al. [4] specify the areas, tools and challenges of MBT approaches. Authors classify existing approaches by (a) SUT covering both software paradigm and domain, (b) test objectives including artifact, level and type, (c) model specification such as language type, model scope, paradigm and characteristics, (d) test generation and (e) test execution. This taxonomy follows the 5-step model mentioned above; however, the division of each step into sub-criterion is close enough to more specific and widely-used [19], at least partially addressed in the most MBT-related articles. Therefore, let us turn to this study, proposing taxonomy to divide approaches or to set requirements developing your own. This is seven-dimensional taxonomy [19] (Table 2, columns 1–3) dividing MBT into: (1) subject, (2) redundancy, (3) characteristics, (4) paradigm, (5) test selection criteria, (6) technology, (7) on/offline (for their description, please see [19]). The first four dimensions are all related to the model specification, together creating the first phase of the MBT, while the test selection criteria and test specification belong to the 2nd and 3rd phases, i.e. test generation and execution. Our approach is described in Table 2 column 4.

One of the most important points is related to the "dynamics" of model specification. Like most MBT approaches, the proposed one is discrete, although it also proved to be used as a continuous. This has been demonstrated in the e-scooter system [25] as an example of internet of things (IoT), which is the subject of current studies [19].

To sum up, the 5-step model, together with the MBT taxonomy provided, constitute the most general classification of MBT approaches. Thus, together with a brief discussion

Table 2. MBT taxonomy (based on [19]).

Phase	Dimension	Aspect	DQMBT
Model specification	Subject	Environment/SUT	SUT
	Redundancy	Shared test & dev model/Separate test model	Separate test model
	Characteristics	Untimed or timed Deterministic or non-det. Discrete or hybrid or continuous	Untimed Deterministic Hybrid
	Paradigm	Pre-post or input domains/Transition-based/History based/Functional/Operational/Stochastic/Dataflow	Pre-post or input domains Dataflow
Test generation	Test selection criteria	Structural model coverage/Data coverage/Requirements coverage/Test case specification/Random & stochastic/Fault-based	Structural model coverage/Data coverageRequirements coverage Test case specification
	Technology	Random generation/ Search-based algorithm/ Model-checking/ Symbolic execution/ Theorem providing/ Constraint solving	Symbolic execution
Test execution	On/Offline	Online/Offline	Offline

on the factors most affecting the popularity of MBT, RQ1 is answered. Now, let us turn to the DQ-model basics briefly addressed in the following section.

3 Research Methodology

This section deals with a brief description of the algorithm, then turning to the DQ-model. As a result, the RQ2 is planned to be answered, forming an input for the RQ3.

3.1 Algorithm and General Architecture of DQMBT

As is stated in [25] the proposed algorithm uses a DQ-model as a testing model and is able to generate a DQ-complete test set (CTS). Its main steps are provided in Table 3.

Table 3. Data quality model-based testing of information systems: algorithm [25]

#	Step
0	Creating DQ-model covering both, input messages and data retrieved from the DB
1	Data quality conditions defined in flowcharts are expanded in a tree-like format
2	For each tree branch, the condition for its reachability shall be established by means of symbolic execution
3	Resolving the conditions results in tests, containing both, input data values, data objects (DB) content, branch execution results. The obtained test set is a DQ-complete test set that ensures testing of all conditions and the results to be stored in the DB

The general architecture of DQMBT is shown in Fig. 1. An important point here is the division of the model initially presented in [12] in **pre-conditions** and **post-conditions**, depending on the analysis phase – the verification of the data to be entered or the data deployment/ retention in the DB. The basics of both terms involved are in line with

[26], according to which precondition refers to the required state of a test item and its environment prior to test case execution, while post-condition - to the expected state of a test item and its environment at the end of test case execution.

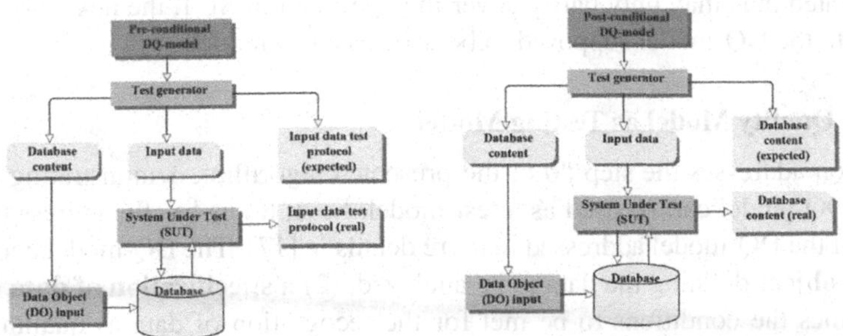

Fig. 1. Pre- and post-conditional testing scheme.

The main actions are performed by a "*Test generator*" using DQ-model to generate test input data, data object content (DB) and protocols – "*Input data test protocol (expected)*" and "*Database content (expected)*". For input data verification, a DQ-complete test set is generated from the pre-defined DQ-model, allowing testing of the quality conditions of the SUT input data for both correct and incorrect data. More specifically, tests are generated that execute all cases of fulfillment and denial of all conditions. We will call these input data checks as *pre-conditional checks.*

The correctness of data retention in the DB is tested after testing the system with a DQ-complete test set which is described by a second DQ-model, called a *post-conditional DQ-model.* The values of data objects stored in the DB to be compared with input data are read and, if the input data are correct, these values should comply with those stored in the DB. Otherwise, i.e. if input data have not been correct, the contents of the DB should not change because the data will not be stored.

The results of the SUT execution are recorded in the "*Input data test protocol (real)*" and the content of the data objects (DB) are read after testing SUT with the generated test input data. The "*Input data test protocol (real)*" must match the "*Input data test protocol (expected)*", although there are possible differences in formatting and texts. If these protocols in general coincide with each other, the SUT is assumed to operate in line with the DQ-model. Otherwise both protocols are sent to developers for further investigation of the reasons for the differences which can indicate errors in the SUT or differences in the DQ-model from programmers programs.

The proposed solution follows [27], which stipulates that "*good testing principles*" suppose that "*part of a test case is a definition of the expected output result*", and "*test cases must be written for invalid and unexpected, as well as valid and expected input conditions*". The proposed testing provides complete testing according to DQ-model, since all quality conditions are tested, reaching full coverage, both by fulfilling and rejecting the conditions. Thereby, a specific test criterion is proposed - *whether the data to be entered are accurately allocated to the data objects (DB) without contradicting the data previously stored.*

The proposed approach is consistent with the "black box" model because information on internal design or implementation of the system is not used. Only the DQ-model is used to generate tests. As the completeness of the testing is highly dependent on the model, i.e. SUT may include activities that are not covered by the DQ-model, and the tests generated thus may only partly cover the operation of SUT, the next subsection is dedicated to the DQ-model supposed to be used as a test model.

3.2 Data Quality Model as Testing Model

This Section addresses the step #0 of the presented algorithm, demonstrating how the proposed DQ-model can be used as a test model. First, let us briefly address the main concepts of the DQ-model addressed in more details in [17]. The DQ-model consists of: (1) **a data object** defining the data to be analyzed, (2) **a specification of data quality**, which defines the conditions to be met for the recognition of data as qualitative, (3) **a quality assessment process** that determines the procedure that must be followed to assess the quality of data. Each DQ-model component is represented by flowchart-like diagram defined in a graphical domain specific language (DSL).

The data object has only those parameters, in which specific user or tester is interested in. This affects performance significantly, since only the parameters which matter are selected for further processing. Data objects of the same structure form data object class that allows processing a collection of data objects in a unified way.

While in the case of data quality analysis it was possible to define a threshold value allowing specifying when the quality of a particular data object is considered to be low, in this study, a threshold value is not defined because each negative result must be inspected reaching a conclusion on the follow-up actions – whether the model should be adjusted to the SUT, either an error has been identified in its operation.

The number of data objects involved is not limited and their number depends mainly on the specific case – the structure of IS and the scope of testing. Both input messages and the DB are considered and treated as data objects, establishing relationships between them. For relationships – they can be created only for data input from an input message data object to a DB data object, or contextual dependencies between them. These dependencies are depicted graphically when defining data objects and their more precise definition is done in the quality conditions by means of logical expressions or SQL queries, i.e. the quality specification can be defined informally or formally, but at the last stage all requirements are replaced by executable artifacts. The data quality specification shall also be determined by the user/ tester, depending on the use-case. Then, a DQ-model used to generate tests is obtained.

The proposed approach proved its effectiveness in applying it to real data sets, analyzing their quality demonstrated in the series of articles. Now let us discuss the DQ-model in more detail by an example.

Defining Data Object. An example of definition of pre- and post-conditions will be demonstrated by looking at the **insertion/** input of insurance reimbursement claims to the insurance system. In a real-life, after an accident the insurance policy holder, i.e. the client, applies to the insurance company to apply for the insurance case. Therefore, the *Client* data object shown in Fig. 2 is defined containing the *Client_id* that describes each customer.

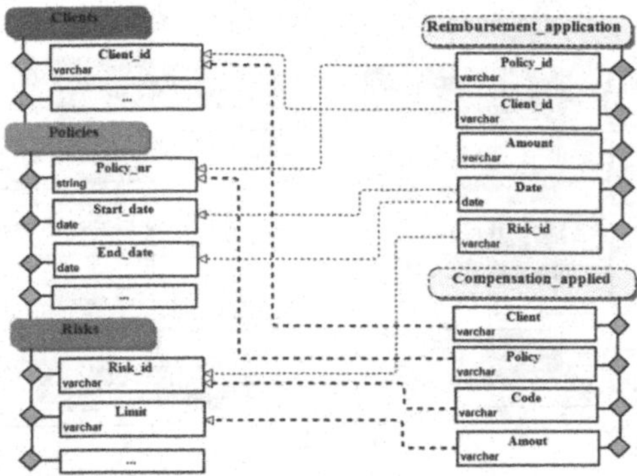

Fig. 2. Definition of the "Compensation application" data object.

Each instance of the *Client* contains a *Policy* data sub-object that contains all the data on the policies purchased by the client - the policy number *Policy_nr* and the policy period, the *Start_date* and *End_date* dates. Each instance of the *Policy* data object contains the *Risks* sub-object, where each instance contains data about the policy's insured risk *Risk_id* and the limits on reimbursement costs. The *Reimbursement_application* data object contains data that are entered by the customer when the compensation is applied. The *Compensation_applied* data object contains data about the registered compensation application that is stored in the DB.

Defining Quality Requirements. The checks for data objects to be entered are described by logical expressions. The operands of logical expressions are values for attributes of data objects, but logical operations are operations allowed in the programming language. For a simple illustration of a proposed idea, operations and conditions can be defined in a pseudo code, but pre- and post-conditions using the SQL. The attributes of each instance of a data object class are checked by these conditions.

The correctness of the data to be entered is controlled by pre-conditions prior to actual data entry into the DB. At the moment of data input it checks whether the values of the attributes of data objects to be entered correspond to the syntax of attribute values and the values of the attributes of data objects correspond to the values of other data objects. Whenever any of the values of the data to be entered do not meet the requirements, further data verification shall be suspended, the owner of the data to be entered shall receive a statement of non-compliance and may make corrections. Figure 3 shows the verification of the correctness of the data to be entered in the case of the claim of reimbursement with pre-conditions established to control the data to be entered using a pseudo code. It is checked successively whether the customer to be entered exists in the covered system or whether the client concerned is the holder of the policy to be entered. When this is the case, it shall be verified whether the claimed reimbursement event occurred during the period of activity of the policy etc.

When the data set to enter has passed a check, the data are entered into the insurance system. After the data are entered, the data recorded in the DB are verified. This process supposes making sure that the data entered are stored correctly in the DB, i.e., the

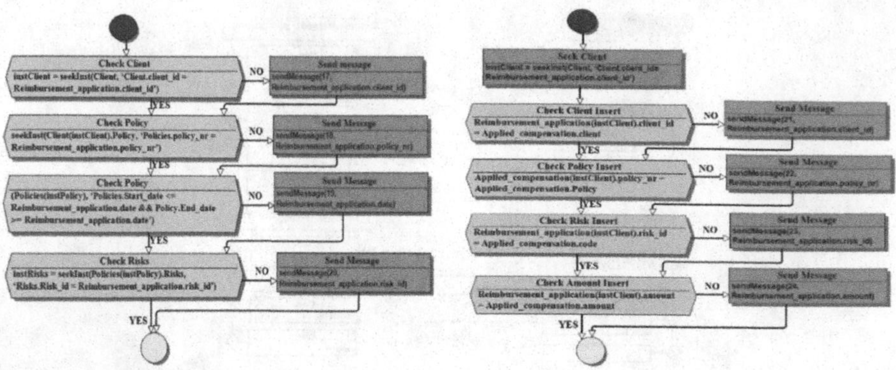

Fig. 3. Pre-conditions for evaluating the DO. **Fig. 4.** Post-conditions for evaluating the DO.

corresponding values of the instance attributes of the *Applied_compensation* data object match the values of the *Reimbursement_application* data object attributes. In the case of an example, the verification of the *Applied_Compensation* data object compares whether the data entered in the claim are stored correctly in the corresponding data item (Fig. 4). The attributes of the *Applied_Compensation* must contain data about the customer *Client*, the policy number *Policy*, the *Codes* of the risks to be reimbursed, and the *Amount* of compensation to be paid. If one of the expected values is not present at the time of fulfilling the conditions, the user shall be informed. This allows verifying the correctness of the system from the point of view of data storage. It should be acknowledged that the example given does not fully reflect the business process, since data objects and conditions can be more complex.

In the light of abovementioned we argue that DQ-model is appropriate for model-based testing, i.e. we are able to cover all the cases we are interested in. The following section will test this hypothesis, by answering the RQ3, i.e. *is the proposed data quality model-based testing suitable for real system testing?*

4 Applying DQMBT to the Insurance Policy System

In this section, we provide a brief insight into existing development and quality assurance processes of an insurance company, following by a description of the part of the SUT. Then we pay attention to the proposal and to the structure of the conditions. This will be followed by a summary of how test data are generated on the basis of established conditions and a follow-up test that serves as a result of the final test.

4.1 Description of the System Under Test

The addressed insurance company has an IT department consisting of several development groups. Each group consists of an analyst, developers and one or two testers. There is an independent group involved in automating tests. Testing is divided into (1) unit testing conducted by developers, (2) functional testing carried out by testers. The functionality introduced is tested with manually prepared tests. The generation of tests is based on the knowledge, experience and functional requirements formed by the systems analyst. To monitor the stability of the software being developed, the automatic regression tests are designed and maintained in Gherkin. Errors found are logged, linking them to the required functionality. In addition, the business requirements specifications, describing the requirements for the entire system, are recorded.

The proposed test method is applied for one of the largest systems - the sales system that ensures the purchase of all insurance policies offered by the company in the Baltic States. It is considered to be the core system, as other systems merely complement this system. It is also one of the most long-standing systems in the company. The system is broken down by insurance product. In the example, this testing method is applied to a housing insurance product. The purchase of a housing insurance product requires the completion of 17 fields divided in 3 main areas: (1) **policy holder and general information** containing policy holder data, contact details, policy period and reward payment arrangement; (2) **insurance object** containing data about insured real estate, such as address, type of property, area, year of construction, etc.; (3) **insurance coverage block** containing data about insurance coverage.

Defining Pre- and Post-conditions: Pre- and post-conditions are defined using SQL which was found to be one of the most appropriate for those tasks [9]. In addition, the SUT already uses a relational DB. The creation of conditions can be divided into 3 steps: (1) define query variables; (2) set up conditions for defined variables; (3) set up conditions for the values stored in the DB. Let's examine these steps in more details.

The 1st step ("define query variables") can be considered as a step in defining the pre-conditions, since the variables are not only assigned names but also defined the expected data type of variables. Thus, for each attribute of a data object, the definition of conditions starts with the determination of its permissible data types.

Once variables have been defined, it is time to focus on creating conditions for attributes of data objects (2nd step). The establishment of statements shall be based on the requirements of the specification of the housing insurance product, in particular the conditions for the fields to be entered. Technical requirements and relationships between the defined fields should also be taken into account. Figure 5 shows 12 quality pre-conditions for the created data object attributes. These conditions are expressed by SQL queries and suppose the sequential execution of the checks and messages to be sent to the user in case of an input mismatch.

Data on a policy holder can be queried on the system in two different steps either (1) by reading an existing client from the DB, either (2) by entering and creating a new client. Because creating a new client is a different, independent functionality and is not directly linked to filling in the data for the housing insurance product policy under consideration, this option will not be addressed. Therefore, the restrictions on the variable *policyHolderCode* are imposed as follows: the private or legal personal code entered exists in the DB and the entry is active, i.e. not closed.

The pre-conditions determine, if the value to be entered is not matched in the DB, no further verification of the data will take place because the policy cannot be filled in without the policy holder. Let us cover a few more examples:

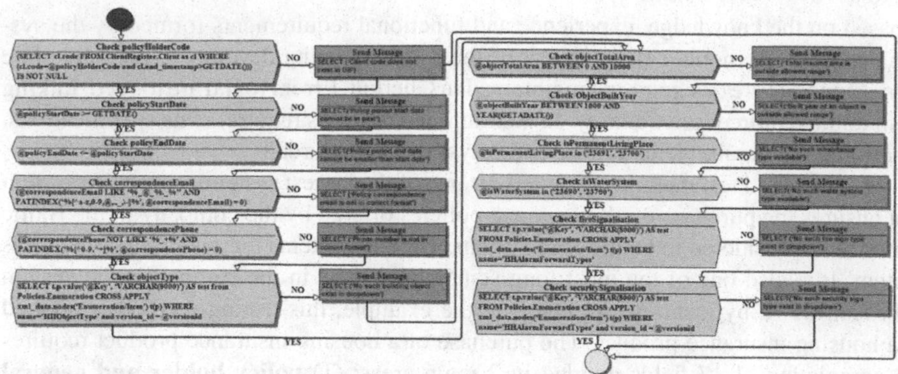

Fig. 5. Pre-conditions for quality requirements for the "Policy" data object.

- *policyStartDate* and *policyEndDate* are linked to each other. The date of entry into force of the policy (1) *policyStartDate* must not be in the past with regard to the time when the policy was formed and issued, and (2) the date of the end of the policy (*policyEndDate*) cannot be earlier than *policyStartDate*;
- when the policy is prepared, the user can choose the policy payment plan and the frequency of payments. The input value chosen should correspond to a value from the defined list. This is also the case for *objectType*, *objectBuildingMaterial* and *isPermanentLivingPlace;*
- the conditions for variables *correspondenceEmail* and *correspondencePhone* require values to meet predefined patterns.

A corresponding error message is arranged for each check if the data to be entered do not meet the expected requirements. In this case, the user may manage the intended input data in the system and observe the behavior of the system in the case of incorrect data, or rectify the intended input data and re-execute the pre-conditions.

Once the conditions are established, input data are prepared and entered into the system, the verification of the correctness of storing the data entered, must be carried out by mapping to them post-conditions – 3rd step took place.

Following the data entry into the system, i.e. the establishment of an insurance policy, a new data object is created in the DB - *Policy*. A sub-object has formed for the *Policy* data object – *Object* to be insured. The policy is characterized by a unique number that will be used to link data values stored in the DB to the policy. Figure 6 provides the conditions for evaluating the *Policy* data object and *Object* sub-object.

Post-conditioning is a lighter process. In general, it is as a verification of the compliance of the data entered in the DB for the relevant DB unit. The post-conditions are defined for 13 DB values including: (1) identifier of the client associated with the policy corresponds to the identification of the value *policyHolderCode* in the customer table; (2) the e-mail address of the contact stored by the policy corresponds to *correspondenceEmail*; (3) the year of construction of the insured object corresponds to the value *objectBuiltYear*; (4) the type of water supply corresponds to *isWaterSystem*; (5) the fire alarm value corresponds to *fireSigna-lisation*, (6) the value of the security alarm

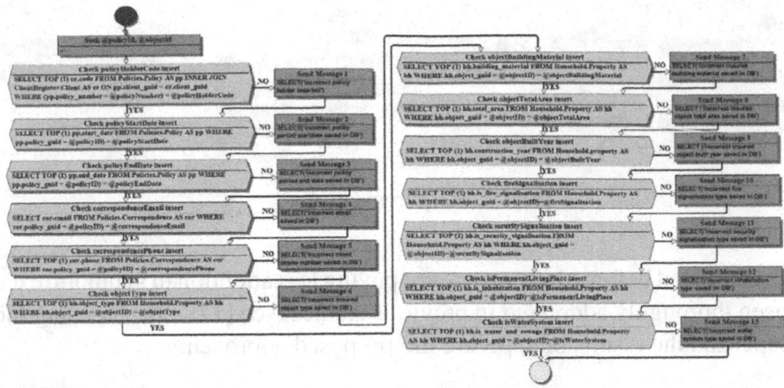

Fig. 6. Post-conditions for quality requirements for the "Policy" data object.

corresponds to the *securitySignalisation* etc. Among other conditions, the checking of the policy holder in the DB differs most, because there is no private or legal personal code for each policy data object, but its unique identifier.

Also in the case of post-condition, an appropriate error message has been mapped for the verification of each variable condition if the data stored in the DB does not meet the expected requirements, are incomplete or cannot be found. In this case, the user shall receive the notification and may draw conclusions on the test example.

After the establishment of pre-conditions and post-conditions, system testing may be performed by first generating test data and validating the test using conditions, then performing the tests in the system and verifying that the final result is stored in the DB using the established conditions.

4.2 Expanding DQ-Model into a Tree and Testing Process

In the proposed algorithm, the first step requires the deployment of a condition model (Figs. 5 and 6) into a tree-like chart given in the Fig. 7. Branch vertices contain numbers, i.e. identifier further used when generating the test, ranging from 1 to 25 in the given example. The tree contains only one node that represents correct data processing from syntactic and context checks to correct data allocation in the DB - branch #1. The branch #2 represents an input data context violation because DB does not contain data about the policy holder in question, while branches #3–#13 point to other violations of the same nature. Branches #14 to #25 indicate incorrect data distribution in the DB. In this example, the branches are linear, however, this is not a mandatory requirement and they may have more complex structure if needed.

In the second phase of the algorithm, the conditions for the realization of the corresponding branches are established using the symbolic execution of the DQ-model conditions. When resolving the conditions for the branch realization in all 25 cases, the test input data are obtained and the content of the data objects (DB). Each instance serves as test input data to complete one of the 25 branches. In other words, the generated test set is a complete DQ-test set. Execution of the SUT with all these tests will achieve the complete testing of the SUT according to the DQ-model criterion.

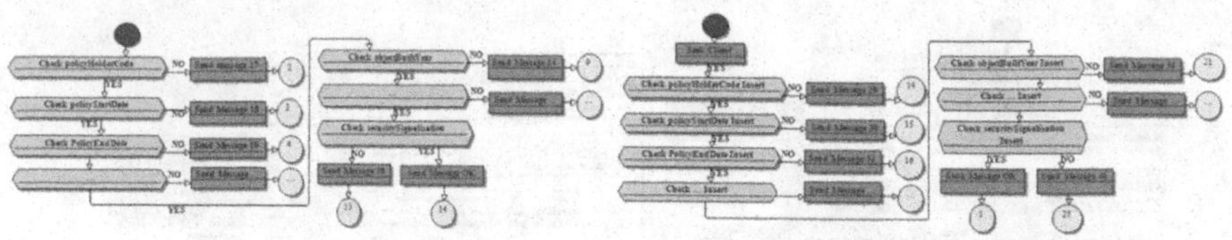

Fig. 7. Requirements tree.

As we mentioned before, this time we are paying less attention to symbolic execution as it has been thoroughly addressed in previous studies (see [12, 25]), focusing more on new concepts and the results of applying the proposed approach.

4.3 Testing Process

When a DQ-model was prepared, tests for complete testing were generated by developed symbolic interpreter. Then, the values of generated data objects are sent to the DB that is done by separate procedure individual for a particular system. This is followed by SUT testing with the DQ tests. The results of the SUT execution must be consistent with the previously obtained protocols. In the case of differences, the inconsistencies between the operation of the SUT and the DQ-model are identified. This can be caused by both errors in the SUT or errors in the specification – the DQ-model. To automate the testing process, most test support tools support the SUT execution with a user-selected set of tests [1, 28–30]. These tests usually are accumulated gradually using the same test support tools. As a result, in most cases the tests records formats are internal formats of these tools which are not related to the tested programs. In the proposed case, the situation is more complicated because all generated tests must be able to be performed automatically in one session. This is achieved by preparing test drivers that establish DB content, calls the cyclic execution of the SUT with all generated tests, and read the DB content after completion of the tests [12].

Since the selected component had long been in industrial operation, it was assumed that, as a carried-out task is a re-testing activity, no defects should be detected. Therefore, a hypothesis on the suitability of the proposed DQMBT to testing a real system was established, i.e. whether any defects will be detected?

5 Research Findings

Generating Test Data and Testing SUT. This section describes the generation of test data to be entered by means of the pre-defined conditions and their use for testing.

The test process starts with generating input data. On the basis of the defined conditions, a total of 66 input data combinations were established, including data values that meet conditions and do not comply with the conditions. Each combination of input data values deals with a certain variable of input data and changes its value according to or inconsistent conditions, with the values of the other input data being recorded, with only the correct ones. Table 4 provides the summary of the total number of tests per attribute, the number of tests covering positive and negative cases, as well as the result. Let us focus on a few of them in more detail.

Table 4. Test data for SUT and its testing results.

Variable name	Condition	Total	Correct	Incorrect	Result
policyHolderCode	Code, exist in DB	4	2	2	OK
policyStartDate	Date, compared to the current date, tomorrow and *n* days in the future	4	3	1	OK
policyEndDate	Date, compared to *policyStartDate* - +*n* days, the same, *n* days before	3	2	1	OK
correspondenceEmail	Pattern	5	3	2	ERROR!
correspondencePhone	Pattern	6	3	3	ERROR!
objectType	List of 13 identifiers	14	13	1	OK
objectBuildingMaterial, securitySignalisation, fireSignalisation	List of 3 identifiers	4	3	1	OK
objectTotalArea	Range	7	4	3	ERROR!
objectBuiltYear	Range	5	2	3	ERROR!
isPermanentLivingPlace, isWaterSystem	Identifier	3	1	2	OK

In the case of *policyHolderCode*, 2 conditions are generated, i.e. (1) physical and legal personal codes that exist in the DB and (2) codes are valid. In contrast, two non-compliance cases where one is a non-existent client in the DB, the other client in the DB with a closed entry, were defined. In fulfilling the preconditions against these examples of input data, the result corresponds to the preconditions in the first two cases expected, in the remaining cases the preconditions for the value of *policyHolderCode* are outstanding and the user receives an error message.

Four values have been defined for the variable *policyStartDate*. Three of four tests are designed according to the conditions on the main data, where the starting date is today, tomorrow and *n* days in the future. One of the tests contains a value that does not meet the conditions and is in the past. The conditions for input data validate the relevant tests for expected results. Similar checks were carried out for the variable *policyEndDate*. No errors were detected because the system is operating as expected.

On the basis of the conditions, the value of the variable *objectType* must correspond to one of the 13 allowable values in the list, identifiers. Here are 14 test examples, including 13 corresponding values and one non-matching value. Similar checks were carried out for the variables *objectBuildingMaterial, isPermanentLivingPlace, isWaterSystem, fireSignalisation, securitySignalisation.* The result of fulfilling the conditions on the input data created coincides with the results originally expected.

For the *objectTotalArea*, 7 tests have been set up, 3 of which are non-compliant with conditions. The values defined shall include limit values corresponding to the specified range. One of the non-matching values is a character that does not correspond to the defined variable format, so that input data cannot be verified with preconditions. Since the definition of the variable type was included in the set of preconditions, it may nevertheless be considered that it has not been fulfilled.

After fulfilling the pre-conditions on the generated input data, it can be concluded that there are problems if the input data to be verified under conditions do not match the expected data type of the value defined in the SQL query. In this case, the user does not receive an appropriate error message but a system error message. This means, that improvements could be included here.

Post-conditional Verification of Data Stored in the Database. After setting up and testing the input data tests, the tests resulting from conditional testing may be inputted to the SUT. For this purpose, automatic tests were used to automatically enter the large number of tests into the system when creating policy forms. At the end of the test, the unique numbers of the policies were created and stored, so that the values of the attributes of the data objects to be checked in the DB can be identified. Because both tests are entered into the system, where the expected result should be positive, and tests where the expected behavior in the system is not expected because of incorrect data values, incorrect data were entered manually in the system.

A total of 66 tests have been completed, where 44 of them have been found to be relevant and 22 of them do not meet conditions. When the correct test example is entered, in all cases, the data objects are stored in the DB with corresponding values entered in the system. If a policy holder to be entered is not found in the DB, the system cannot continue to fill in further data, nor does the system generate a unique identifier. In this case, the data objects are not created, so it is not possible to verify data stored in the DB. The results of incorrect outcomes that have revealed the incorrect system behavior are summarized in Table 5. Otherwise, the results vary according to the variable type to be entered. In some cases, the user interface displays a warning about the conditions for non-compliance with the data, but the value is stored in the DB, whereby the conditions on the values stored in the DB are met.

Table 5. Testing results.

#	Variable	Expected behavior	Real behavior
8	*policyStartDate*	A warning about an **incorrect format**	**Error message #2 #3**
11	*policyEndDate*		
13, 14	*correspondenceEmail*	A warning about an **incorrect format**	**Incorrect value stored in DB**
18-20	*correspondencePhone*		
43, 44	*objectTotalArea*	A warning about **a value outside the interval**	**Incorrect value stored in DB**
48, 49	*objectBuiltYear*		

If an inappropriate date is set up, the user receives a warning but the date is automatically replaced by the default value. As regards the *correspondenceEmail* and *correspondencePhone*, after the data are entered, the user receives a message about the format that does not match the field to be entered, but the follow-up to the data stored in DB is a successful result, i.e. the incorrect value has been stored in the DB. This is also the case for *objectBuiltYear* and *objectTotalArea,* where values outside the interval allowed. These are potentially dangerous situations that have not been disclosed by previous testing methods.

The incorrect input of variables where the values are selected for the predefined list was not possible, as it includes only the values specified in the system. This is also the case for the input taking place by checkboxing. For wider discussion see [31].

Thus, given that the defects not disclosed by previous testing methods during the operation and testing of the system over several years, were revealed, we claim that DQMBT has proved its suitability for real system testing. However, it must be recognized, it has limitations and cannot be used as the only test method.

6 Discussion on Findings

The testing technique proposed is tested on a system that is one of the most long-standing systems in the company. Given this and the fact that we believe that a better use of this method would be by testing the systems being at the development stage, a number of system deficiencies were found as a result of testing. By applying the new method to the SUT, we conclude that (1) based on the input data values created during the system specification, it is convenient to prepare tests for both positive and negative scenarios; (2) verification of the values stored in the DB can serve as a kind of system security check by means of conditions, i.e. when entering data into the system and verifying data stored in the DB under conditions, it becomes possible to determine whether the wrong value is stored in the DB.

However, (1) formalizing system requirements with SQL queries can sometimes be difficult. This is the case when the requirements for the permissible values for input fields in the system are consistent with the configuration installed in the system or other system conditions. Too much flexibility complicates the development of conditions relating to the data to be entered, and the setting-up of conditions must have a more in-depth knowledge of the design of the system and system-related DB. However, when the requirements for data to be entered into the system are simple, formalized conditions are also simple. (2) In the case of restrictions on the type of data for the values to be entered and when the user wants to control the input data to be tested that contains a data type that does not match the value, SQL is unable to process this information and the user receives a system error instead of a non-compliance error - here we could look for a solution for improvements; (3) since the entry of the test data is facilitated by the establishment and use of the automatic tests, the most time is spent on generating the test data and preparing the test set for pre-conditional testing.

Another point is related to defined models and their accuracy over time. As one of the features of software development is its variability, it is clear that expected requirements and specification of requirements may change during the development process. As a result of these changes, the established pre-condition set must be maintained regularly. It should also be taken into account that the prepared set of input data to be verified can be updated by the user, that results in a necessity to verify the correctness of the attributes of the data objects.

Verification of the post-conditions should be carried out every time a test case is executed, but it is essential to take into account that the SUT is always subject to changes. Changes may be caused by another user of the SUT or software that is integrated with it. Even if they are not supposed to affect the data stored as a result of the execution of these tests, it is always possible that the system with which it is linked works erroneously. For this reason, it is desirable to re-execute the conditions on the data stored in the testing DB, regardless of the previous result of the execution of pre-conditions. In this way, the correctness of the data may be monitored.

It is also clear that the demonstrated example is simple enough as there are quite precise rules that are reused while defining modeling conditions. However, the proposed approach can be used for testing more complex systems with less detailed rules, as was the case for e-scooters [25]. The only difference is that a modeling task with the definition of conditions may take longer to reach an accurate and sufficiently detailed model, while other aspects of the approach will still be the same.

And the last point is that the proposed criterion does not guarantee the detection of all errors in the operation of the SUT. For example, an SUT operation that record data in inappropriate DB locations is not controlled. Therefore, we conclude that the proposed DQMBT provides improvements to existing testing processes, but is not sufficient for full testing of the system being used as a complementary and can work well in combination with other testing techniques.

7 Conclusions

Existing testing techniques do not guarantee the development of high quality and reliable software, as well as full testing for all possible conditions under which the system can be used. This can be achieved, at least in part, through systematic testing, for example by means of model-based testing addressed in the paper.

The study has answered 3 research questions raised. First, the success factors of MBT were identified, explaining its popularity. It was then determined how existing approaches are classified. Then, the appropriateness of the data quality model was verified, presenting an alternative data quality model-based testing approach - DQMBT. DQMBT's suitability for testing of real systems was analyzed by applying it to the system of insurance applications. While following the classical rules intended for MBT, the proposed approach covers only part of the functional testing of IS, more precisely complete testing of the input data and the verification of the relevance of data stored in the DB with input data. This limitation, however, facilitates the creation of CTS testing the selected areas of IS thoroughly. This is supported by the use of DQ-model, which allowed achieving the needed characteristics without significant complication, as is the case for OCL and UML. This would significantly improve the overall quality of IS, which is today one of the most important challenges [10, 11].

In addition, the proposed solution does not meet the limitations regarding the application area, as is the case for the most MBT approaches [5]. Moreover, it was already demonstrated in the e-scooter system [25], thus showing the model belongs to both discrete and continuous ones, which is the subject of current studies [19].

Further studies on the topic include more extensive testing of the proposed approach, which should result in a quantitative comparison with other strategies. The issue on how to deal with system events should be addressed when one event takes place more quickly than others, thus affecting the result of previous events. In addition, the options to automate the testing process is one of the key aspects to be covered in future studies, including the possibility of using this testing approach more easily for a wider range of users, thereby increasing its commercial potential.

Acknowledgements. This work has been supported by University of Latvia project AAP2016/B032 "Innovative information technologies".

References

1. Garousi, V., Elberzhager, F.: Test automation: not just for test execution. IEEE Softw. **34**(2), 90–96 (2017). https://doi.org/10.1109/MS.2017.34
2. Saeed, A., Ab Hamid, S.H., Mustafa, M.B.: The experimental applications of search-based techniques for model-based testing: taxonomy and systematic literature review. Appl. Soft Comput. **49**, 1094–1117 (2016)
3. Villalobos Arias, L., Quesada López, C., Martínez Porras, A., Jenkins Coronas, M.: A tertiary study on model-based testing areas, tools and challenges: preliminary results (2018)
4. Uzun, B., Tekinerdogan, B.: Model-driven architecture based testing: a systematic literature review. Inf. Softw. Technol. **102**, 30–48 (2018)
5. Gurbuz, H.G., Tekinerdogan, B.: Model-based testing for software safety: a systematic mapping study. Softw. Qual. J. **26**(4), 1327–1372 (2018)
6. Iqbal, M.Z., Sherin, S.: Empirical studies omit reporting necessary details: a systematic literature review of reporting quality in model based testing. Comput. Stand. Interfaces **55**, 156–170 (2018)
7. Nikiforova, A., Bicevskis, J.: Towards a business process model-based testing of information systems functionality. In: Proceedings of the 22nd International Conference on Enterprise Information Systems - Volume 2: ICEIS, pp. 322–329 (2020). ISBN: 978-989-758-423-7. http://dx.doi.org/10.5220/0009459703220329
8. Perez-Castillo, R., Carretero, A.G., Rodriguez, M., Caballero, I., Piattini, M. Mate, A., et al.: Data quality best practices in IoT environments. In: 2018 11th International Conference on the Quality of Information and Communications Technology (QUATIC), pp. 272–275. IEEE (2018). https://doi.org/10.1109/QUATIC.2018.00048
9. de Cleva Farto, G., Endo, A.T.: Evaluating the model-based testing approach in the context of mobile applications. Electron. Notes Theor. Comput. Sci. **314**, 3–21 (2015)
10. Ziemba, E., Papaj, T., Descours, D.: Assessing the quality of e-government portals-the polish experience. In: 2014 Federated Conference on Computer Science and Information Systems, pp. 1259–1267. IEEE (2014). http://dx.doi.org/10.15439/2014F121
11. Karabegovic, A., Ponjavic, M.: Geoportal as decision support system with spatial data warehouse. In: 2012 Federated Conference on Computer Science and Information Systems (FedCSIS), pp. 915–918. IEEE (2012)
12. Bicevskis, J., Bicevska, Z., Nikiforova, A., Oditis, I.: Data quality model-based testing of information systems. In: 2020 15th Conference on Computer Science and Information Systems (FedCSIS), pp. 595–602. IEEE (2020). https://doi.org/10.15439/2020f25
13. Mohacsi, S., Felderer, M., Beer, A.: Estimating the cost and benefit of model-based testing: a decision support procedure for the application of model-based testing in industry. In: 2015 41st Euromicro Conference on Software Engineering and Advanced Applications, pp. 382–389. IEEE (2015)

14. Schieferdecker, I.: Model-based testing. IEEE Softw. **29**(1), 14 (2012)
15. Utting, M., Legeard, B.: Practical Model-Based Testing: A Tools Approach. Elsevier, Amsterdam (2010)
16. Jorgensen, P.C.: Software Testing: A Craftsman's Approach. CRC Press, Boca Raton (2018)
17. Nikiforova, A., Bicevskis, J., Bicevska, Z., Oditis, I.: User-Oriented Approach to Data Quality Evaluation. Journal of Universal Computer Science **26**(1), 107–126 (2020)
18. Muniz, L., Netto, U.S., Maia, P.H.M.: A Model-based Testing Tool for Functional and Statistical Testing. In Proceedings of the 17th International Conference on Enterprise Information Systems (ICEIS), pp. 404–411 (2015)
19. Utting, M., Pretschner, A., Legeard, B.: A taxonomy of model-based testing approaches. Softw. Test. Verification Reliab. **22**(5), 297–312 (2012)
20. Zander, J., Schieferdecker, I., Mosterman, P.J.: A Taxonomy of Model-Based Testing for Embedded Systems from Multiple Industry Domains, vol. 1, pp. 3–17. Taylor & Francis Group:CRC Press, Boca Raton (2011)
21. Dias Neto, A.C., Subramanyan, R., Vieira, M., Travassos, G.H.: A survey on model-based testing approaches: a systematic review. In: 1st ACM International Workshop on Empirical Assessment of Software Engineering Languages and Technologies, pp. 31–36 (2007)
22. Guerra, E., Soeken, M.: Specification-driven model transformation testing. Softw. Syst. Model. **14**(2), 623–644 (2015)
23. Carvalho, G., Barros, F., Lapschies, F., Schulze, U., Peleska, J.: Model-based testing from controlled natural language requirements. In: Artho, C., Ölveczky, P. (eds.) Formal Techniques for Safety-Critical Systems. Communications in Computer and Information Science, vol. 419, pp. 19–35. Springer, Cham (2013). https://doi.org/10.1007/978-3-319-05416-2_3
24. Hierons, R.M., Bogdanov, K., Bowen, J.P., Cleaveland, R., Derrick, J., Dick, J., et al.: Using formal specifications to support testing. ACM Comput. Surv. **41**(2), 1–76 (2009)
25. Nikiforova, A., Bicevskis, J., Bicevska, Z., Oditis, I.: Data quality model-based testing of information systems: the use-case of E-scooters. In: Proceeding of 7th IEEE International Conference on Internet of Things: Systems, Management and Security (2020). https://doi.org/10.1109/IOTSMS52051.2020.9340228
26. ISTQB. - https://www.istqb.org/downloads/send/20-istqb-glossary/210-istqb-glossary-3-2-ga-release-notes-final.html
27. Lewis, W.E.: Software Testing and Continuous Quality Improvement. CRC Press, England, UK (2017)
28. Rafi, D. ., Moses, K.R.K., Petersen, K., Mäntylä, M.V.: Benefits and limitations of automated software testing: systematic literature review and practitioner survey. In: 2012 7th International Workshop on Automation of Software Test (AST), pp. 36–42. IEEE (2012)
29. Loyola, P., Staats, M., Ko, I.Y., Rothermel, G.: Dodona: automated oracle data set selection. In: Proceedings of the 2014 International Symposium on Software Testing and Analysis, pp. 193–203 (2014) http://dx.doi.org/10.1145/2610384.2610408
30. Kaur, H., Gupta, G.: Comparative study of automated testing tools: selenium, quick test professional and test complete. Int. J. Eng. Res. Appl. **3**(5), 1739–1743 (2013)
31. Capa, V.: Formalizētas specifikācijas vadīta testēšana. Formalized specification-based testing (in Latvian) (2020)

Advanced Comprehension Analysis Using Code Puzzle

Considering the Programming Thinking Ability

Hiroki Ito[1(✉)], Hiromitsu Shimakawa[2], and Fumiko Harada[3]

[1] Graduate School of Information Science and Engineering, Ritsumeikan University, Kyoto, Japan
hirokiito6900@de.is.ritsumei.ac.jp

[2] College of Information Science and Engineering, Ritsumeikan University, Kyoto, Japan
simakawa@cs.ritsumei.ac.jp

[3] Connect Dot Ltd., Kyoto, Japan
harada@de.is.ritsumei.ac.jp

Abstract. In programming education, the instructor tries to find out the learner who needs help by grasping the understanding using a written test and e-learning. However, in reality, not many learners will acquire the skill of writing source codes. This kind of current situation implies that programming ability of learners cannot be measured by tests that require knowledge. This paper focuses on not only the knowledge items required for programming but also the programming thinking (computational thinking), which is the ability to combine the constituent elements of the program. In this paper, we propose a method to estimate the learner's understanding from the learner's process to solve the code puzzles that require programming thinking as well as knowledge. The experimental result with the interface showed that the proposed method could estimate with the accuracy of 80% or more. The accurate measurement of the learner's programming ability contributes to developing the learner's true programming ability, which cannot measured by only the score of written tests. In addition, the importance of each variable in the behavior analysis leads to the identification of learner's misunderstanding factors and the improvement of class contents. This study shows that it is possible to estimate the comprehension level of a programming thinking ability from only behavior of code puzzle, without sensors. The ability of learners to actually write programs is more important than their grades. This research can be developed to help develop the Information Technology talent we need in this era.

Keywords: Programming education · Learning analytics · Programming thinking ability · Code puzzle · Program structure · Factor inference

This work was not supported by any organization.

E. Ziemba and W. Chmielarz (Eds.): ISM 2020/FedCSIS-IST 2020, LNBIP 413, pp. 45–64, 2021.
https://doi.org/10.1007/978-3-030-71846-6_3

1 Introduction

According to the modern research trends in the field of programming education, there are many studies that predict the performance and the learners who drop out [1]. Our university is also using e-learning to grasp the status of understanding. However, there are many learners who cannot write sources in spite of passing written tests that ask their knowledge about grammar and how to express algorithms. The programming ability cannot be measured only by verifying only learner's knowledge. This is because programming skill also requires the ability to construct program elements logically with a perspective [2,3]. Nesbit *et al.* [4] also speculates that ITS within Computer Science and Software Engineering Education contain a higher proportion of procedural learning goals, and a lower proportion of conceptual learning goals. In the actual situation, the only way to verify the true programming ability of a learner, which consists of both knowledge and programming thinking ability, is for the instructor to stand next to the learner and watch the answer. Most of the current educational settings use measurement methods that are biased in terms of knowledge because it is easy to grasp the understanding situation. As the result, many learners will not be able to understand the intention of the task and to acquire the ability to realize it. From such the situation, we explored a research question: How effective is the method that can estimate the learner's understanding situation by considering programming thinking ability.

This paper discusses an understanding analysis focusing on programming thinking ability. Programming thinking ability means "what kind of combination of movements is necessary to realize a series of activities intended by oneself, and how to combine symbols corresponding to each movement. And the ability to logically consider how to improve the combination of symbols to get closer to the intended activity" [5].

This paper defines the learning item achievement level as the degree of acquirement of the knowledge given in a lecture, materials, and so on. It is a contrasting skill of the programming thinking ability [1]. Most current programming education support focuses on the learning item achievement level. Therefore, there is an urgent need to establish a learning support method that considers programming thinking.

The remaining of the paper is as follows:
Section 2 introduces the current status of related works and the existing knowledge and research used in this study. Section 3 describes the method of comprehension analysis using the behavior of code puzzles in detail. It also describes the feedback to be obtained. Section 4 describes the design of the code puzzle problem and the variables used in the analysis. Following that, the purpose and method of the experiment are explained, and we show the results. Section 5 discusses the results of the experiment and its usefulness. Section 6 describes the conclusions, limitations, and future of this paper.

2 Related Works

2.1 Trends and Specific Examples

According to a recent report on intelligent tutoring systems [6], adaptive feedback has been getting popular in recent years. They give feedback in a variety of ways. The major difference is between those that give step-based hints and those that give summative feedback on the submitted code. It still refers to the inherent difficulty in developing an understanding of the structure of the program and the associated problems. The following are some specific examples of research:

The method proposed by Jadud *et al.* [7] is an early study that identifies learners who need guidance using compilation errors. However, this focuses only on the error, and it is not possible to measure why the error occurred and how much the learner understood.

Mysore *et al.* [8] proposed a Web system Porta that can identify the part where the learner is struggling. The comprehension factor is inherently intricately intertwined. However, Porta estimates the comprehension level only by focusing on the fact that the learner takes time. Therefore, the ability and growth of learners are not taken into consideration. Guo *et al.* [9] proposed an interface that supports one-to-many programming learning in real time and its implementation Codeopticon. However, Codeopticon depends on the quality of the instructor and forces a heavy burden on the instructor. Asai *et al.* [10] identified the cognitive load on learners and the factors that caused the cognitive load by the blank filling assignment. However, the blank-filling assignment prevents the flow of logic. It cannot be said that it considers programming thinking. Many of these existing researches focus only on the aspect of knowledge neglecting the degree of understanding based on logic(programming thinking). As a study focusing on behavior, Ihantola *et al.* [11] estimated the difficulty level of a task using a decision tree from the answering process such as answering time and keystroke of a programming task. They suggested that the answer process brings significant difference in learner's comprehension. However, they did not mention programming thinking ability and cannot estimate factors of misunderstanding.

As one of the traditional programming education formats, there is Parson's programming puzzle proposed by Parsons *et al.* [12]. This is introduced as a tool that is easy for beginners to work on. Parson *et al.* asserted that code puzzles can identify specific points and errors that the learner has stumbled. They argued that, since the code puzzle answer is a well model answer, learners can relive good programming practice. However, although this tool focuses on the acquisition of programming thinking ability, it does not estimate the degree of comprehension.

2.2 Programming Learning by a Code Puzzle

This research uses a code puzzle. A code puzzle is a rearrangement assignment where the learner rearranges code fragments such as source code and pseudo code and assembles them to perform appropriate processing. Parson *et al.* said that, for beginners, code puzzles are more effective and better at nurturing logical thinking than full-coding. Moreover, code puzzles present the logic flow unlike the blank filling assignment. Code puzzles simplifies to acquire a feature of the learner's learning behavior used for estimation of comprehension from the actions of selecting, moving, and rearranging blocks.

2.3 Schema for Programming Tasks

Many works pay attention to schema in cognitive psychology. Schema in cognitive psychology refers to the relationship between thought patterns and knowledge for problem solving in human long-term memory [13]. When a person solves an assignment of programming, he or she reads statements specifying the assignment requirements. He or she builds a plan to constitute program structures and codes embedded among them, to satisfy the requirements. The paper refers to the plan to make a program satisfying the requirements as a perspective. Needless to say, if he/she has no adequate knowledge and thought patterns for programming, it is not possible for him/her to conceive a perspective to write proper codes. We can consider that humans combine knowledge to solve assignments. In the process of combining knowledge, humans attain thought patterns to construct a schema from the knowledge and thought patterns. The integration of knowledge and thought patterns here is exactly what the programming thinking signifies. In this research, we consider to estimate the degree of schema construction from the combination of knowledge and logic based on the schema theory.

3 Educational Support Using Behavior

3.1 Overview of the Method

The purpose of this study is to provide novel education support method focusing on not only the achievement level for learning items as emphasized in ordinal education supports but also the ability to organize those learning items. The latter is referred to as programming thinking in the study. The method proposed in the study encourages novice programmers to assemble an intended program making the best use of those learning items with a specific perspective. The achievement level for learning items in this paper is defined as the amount of learner knowledge necessary to solve given programming assignments. On the other hand, the ability of the programming thinking is defined as the ability to combine the pieces of the knowledge so as to design program codes that meet the requirements specified in the assignments. The ability of programming thinking cannot be measured without observing the programming process, which is the learner's behavior. This is because programming thinking occurs in the process of constituting program codes. Learner who have achieved the programming thinking will constitute program codes satisfying the assignment requirements with specific perspectives in their mind, where they imagine main logic flows in answer codes. Therefore, it is impossible to judge whether a learner has achieve

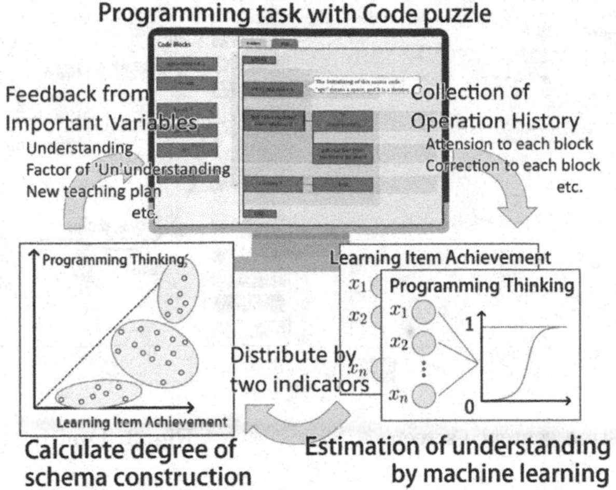

Fig. 1. Schematic diagram of the proposed method

the ability of the programming thinking, unless the learner's behavior is investigated during the answering process.

Since measurement of the ability of programming thinking needs an instructor watches the learner's programming process, code puzzles are promising candidate tools to examine whether learners achieve to combine knowledge on learning items for programming. The method proposed in this research estimates the ability of the programming thinking through analyzing the process in which learners solve code puzzles. The outline of the educational support method aimed at in this research is shown in Fig. 1. First, learners solve the code puzzle. The tool developed for the method using code puzzles is the proprietary application that runs on the WEB. The tool collects the operation logs of the learners. The research assumes operation logs would differ depending on learners. Based on this assumption, the understanding levels of the learner are recognized from collected operation logs using classification based on machine learning. The research adopts a machine learning model accountable for the reason for the classification so that the reason can be presented as feedbacks to both of learners and an instructor.

3.2 Collecting Learner Behavior Using Code Puzzles

The method proposed in this study uses the tools shown in Fig. 2 and 3 to collect learners' characteristic behaviors to measure the understanding level. The tools represents program structures based on the notation of PAD proposed by Futamura *et al.* [14]. In the proposed method, an exemplary program code that satisfies all the requirements given in an assignment is divided into code fragments or pseudo code with functional cohesion. A code fragment generated by the division is referred to as a block in this paper.

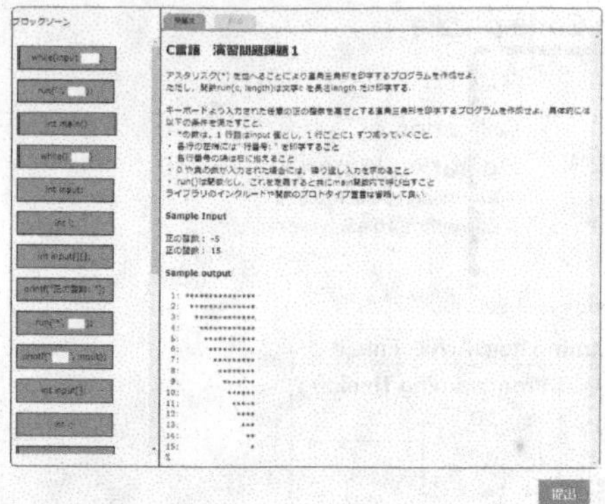

Fig. 2. Question sentence screen

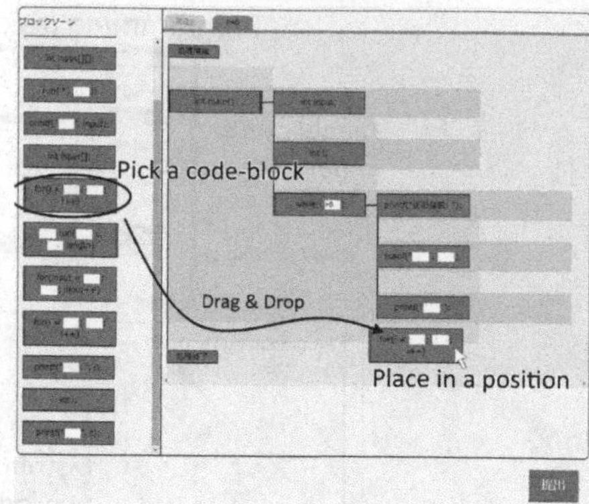

Fig. 3. Drawing screen

As shown in Fig. 2 and 3, The learners assemble the program using the blocks so that all the given requirements are satisfied. This can be regarded as a puzzle that oblige learners to consider the arrangement of blocks so that the requirements are satisfied. The proposed method refers to the arrangement task of blocks as a code puzzle. The proposed method examines the learners' thinking processes during solving the code puzzle in order to judge whether the learners have attained programming thinking or not. When the learners answer with a perspective of how to combine the blocks toward satisfaction of the given requirements, they will place the blocks in the correct position without hesitation. On the contrary, the learners without such a perspective will wonder where to place the blocks. The learners who incorrectly interpreted the meaning of the assignment requirements will quickly move blocks to wrong positions. In this way, it is assumed that the learners' thinking processes during answering appears in the behavior of moving blocks. Based on this idea, the proposed method analyzes the behavior of each learner collected during the learner is placing blocks.

The learner can switch from the assignment sentence screen (Fig. 2) and the drawing screen (Fig. 3) using the tabs at the top of the tool. The learner interprets the assignment from the task sentence as shown in Fig. 2, selects suitable blocks from the blocks displayed on the left side of the tool as shown in Fig. 3, and assembles them by drag and drop. Some of the blocks contain blanks. Blocks with blanks increase the degree of freedom to express procedures in a program. They are introduced to cause learners to get lost. When learners hover their mouse devices over blocks, the description of the blocks is displayed. Learners finish their answering process by pressing the submit button when they believe they have completed constitution of code blocks.

3.3 Explanatory Variables Collected by the Tool

In the proposed method, the understanding level is estimated from the behavior of the learner collected by the tool. Suppose we know the understanding of past learners whose behavior we collected. To classify the behavior of a new learner based on the behavior and the understanding of past learners, the method uses the random method and the logistic regression as machine learning methods. Since the two methods can calculate the importance of variables, it is possible to consider factors that play vital roles in the classification. The random forest method is more accurate than the logistic regression, while the logistic regression let us know the probability of a specific learner to be classified into the class corresponding to learners who attain good understanding. This classification probability can be grasped as an understanding level of 0.0 to 1.0.

According to the schema theory [13], a person combines knowledge to solve tasks. The programming thinking, which is the power to combine knowledge, is considered to be based on knowledge. On the other hand, attainment of only knowledge never leads to improvement of programming skill. The measurement founds on assumption that the real skill for programming cannot be attained unless learners achieve the learning items and the programming thinking in a balanced way [15]. The understanding is measured with two quantities: the achievement level of learning items and the ability of programming thinking. Therefore, the proposed method applies machine learning to create two models, each of which corresponds to the achievement level of learning item and the ability of programming thinking.

The degree of schema construction is defined as follows.

$$d = \begin{Bmatrix} \frac{1}{\sqrt{2}} \\ \frac{1}{\sqrt{2}} \end{Bmatrix} \bullet \begin{Bmatrix} k \\ l \end{Bmatrix}, \quad e = (f(k) - l)^2 = (k - l)^2, \quad s = d - e$$

where k and l stands for the achievement level of learning item and the ability of programming thinking, respectively. We assume k and l range from 0.0 to 1.0.

d is the dot product with the unit vector $(\frac{1}{\sqrt{2}}, \frac{1}{\sqrt{2}})$ and vector $(k.l)$. In other words, d takes the maximum value if both of k and l are 1.0. d become larger as k approaches to l. From the geometrical perspective, as d becomes larger the direction vector $(k.l)$ approaches to that of the line with the slope 45 degrees, whose direction vector is $(\frac{1}{\sqrt{2}}, \frac{1}{\sqrt{2}})$. e is the squared error from the line $f(x) = x$. When calculating s, which is the balance between learning item achievement and programming thinking ability, e indicates the imbalance as a penalty.

3.4 Factors of Misunderstanding for Each Learner

To attain the programming skill, learners should get not only knowledge provided as learning items but also the ability of programming thinking to combine the knowledge. Well-balanced attainment of them is crucial for learners engaging in programming. Each assignment includes the learning items and programming thinking required to solve it. Through solving the assignment, learners understand both of them. In order to estimate the understanding of a learner, it is possible to obtain the classification result using the two machine learning models. Using logistic regression and random forest, it is possible to see which explanatory variables and how much influenced the classification by referring to the regression coefficient and variable importance, respectively. These indexes express the factors that classify learners efficiently. The difference of the behavior of the two classes makes it explicit what factor distinguishes them. The position on the code to which the factor corresponds, that is, the misunderstanding point, becomes clear. In this study, this factor is called the misunderstanding factor. In addition, these indicators provide clues as to where learners tend to stumble and which parts a learner does not know.

3.5 Feedback to Learners and Teachers

The proposed method provides two types of feedback:

- the degree of Schema construction including programming thinking, and
- misunderstanding factors and misunderstanding points that take programming thinking into consideration

The schema construction can visualize understanding of a large number of learners with unified numerical values. Based on the schema construction, the learner can grasp how much he/she understand compared with others. Furthermore, the instructor can detect learners who need guidance at early stage. In addition, when analysis is performed over a period of time, it is possible to calculate the degree of improvement as well as to find stumbling based on changes in the degree of schema construction.

The misunderstanding factors and misunderstanding points can be considered from the coefficient of logistic regression and the importance variables of the random forest. The comparison of the learner's behavior based on the variables of high importance enable us to grasp the misunderstanding factors. This analysis tells the learner what part they do not understand or how they tend to answer. It also informs the instructor of what part confuses learners in addition to what learners fail to understand. Therefore, it is possible to create a new teaching plan for individuals and the whole.

4 Experiment

4.1 Creation of Suitable Assignments for Code Puzzles

The effectiveness of the tool to measure the ability of programming thinking depends on how suitable assignments and code blocks are. In the study, the following strategy is adopted to settle assignments and code blocks. Code puzzles are not suitable for complex assignments. The more complex the assignment, the more code blocks. An excessive number of code blocks would increase the load on the learner's "choice" behavior, which is out of the goal of the proposed method. Therefore, from introductory assignments university programming exercises specified with short sentences, this study selected ones to be solved with as many programming elements as possible. Since this study targets beginners of programming, we chose a relatively simple assignment to be assigned for freshmen. The target language is the C language.

Assignment Sentence

Using a function call, write a program that prints right-angled triangles by lining up asterisk (*) characters. Prototype declaration of the function is "void run(char c, int length);"
That is, the function run() prints the character 'c' for 'length' times. For example, the function can print 4 consecutive asterisks (*) or 5 consecutive blanks. Create a function, run(), and write a program that prints a right-angled triangle whose height is a positive integer n, which is specified from the keyboard. It must meet the following conditions:

- **The number of asterisks should be n in the first line and decreases by 1 in each line.**
- **The line number should be printed on the left-hand side of each line.**
- **The edge of each line should be aligned to the right.**
- **If an integer less than or equal to 0 is entered, the program should prompt a positive integer, throwing away the input.**

Fig. 4. Assignment to print right angle triangle

The 2 assignments are presented along with their execution examples in Fig. 4 and Fig. 5. The assignment in Fig. 4 aims to make novice programmers learn to implement the procedure printing a given character a given number of times in a row using a C language function. The one in Fig. 5 demand them to learn how to make functions returning floating numbers.

4.2 Creation of Suitable Code Blocks for Code Puzzles

The effectiveness of the method to measure the ability of the programming thinking depends on code blocks as well as assignments. In order to analyze puzzlement of learners, the code blocks should cause the learner to get puzzled. However, at the same time, the code block should not impose extra load on the subject's thinking. In order to improve code blocks, we let five learners to solve the two assignments from scratch instead of code puzzles. The actual mistakes that occurred in the actual answering process were reflected in the code blocks of the code puzzle. The errors that actually occurred in this preliminary experiment are listed in Table 1. As the table presents, there are at least 2 kinds of mistakes: low understanding on syntax and no ideas for proper algorithm.

Assignment Sentence

Let us consider the case where the consumption tax goes to 10% but that on food products remains at 8%. To promote cashless purchasing, 2% of the total amount of purchases using a credit card will be returned as points, excluding sales tax. Create a simple POS system to perform this calculation.
We assume that a supermarket offers the following five products.
- Beef 1024 JPY (Food)
- Chinese cabbage 197 JPY (food)
- Apples 125 JPY (food)
- Toothbrush 187 JPY (non-food)
- pot 3050 JPY (non-food)

First, each commodity is distinguished by flag 1 and 0 for food and non-food, respectively. Create function amount_paid() that takes the price of each product and its flag as parameters and returns the paid amount of the product including consumption taxes in a floating-point number. Next, let us distinguish payment methods with credit cards and those with cash using flag 1 and 0, respectively. Create a function amount_of_point() that takes the total price without taxes expressed in an integer and the payment method flag as parameters and returns the amount of the provided point in an integer. These two functions should be defined after the main() function.
Using the functions, write a POS program so that every time the POS reads an item it prints the item price including the consumption tax. After all items are read, the POS program should also print the total amount of payment without consumption tax and with consumption tax, along with the point to be provided. In this program, when the POS read each item, it shows the tax-included price of the item to two decimal places for verification. It calculates the tax-included price of each item, without rounding down the number of decimal places. When printing the total tax-included amount and the points, any number of decimal places should be truncated to make it easier for consumers to know the amount of money he or she should pay and the points to be earned.

Fig. 5. Assignment to calculate points for purchase

Based on the mistakes, we have prepared code blocks. For mistakes regarding to syntax, blanks are embedded in code blocks. The blanks are set in the positions corresponding to Table 1. Learners without enough programming skills are likely to fail in fulfillment of proper codes in those code blocks, even if they choose the code blocks. Learners with no ideas for right algorithm would have problems to place right code blocks on right positions, because they lack in the comprehension in the position where each component of programming should be placed. In terms of incomprehension of the components of programming, we induce the confusion by mixing wrong code blocks similar to the right ones.

4.3 Explanatory Variables Collected by the Tool

The tool analyzes the relationship of the learners' behaviors during answering with their programming ability through machine learning models. In this research, the behavior of the learner is the explanatory variables and the understanding of the learner is the response variables. Before applying this method, we consider what the explanatory variables correspond to the learner's thinking so

Table 1. The errors that actually occurred in preliminary experiment

Assignment in Fig. 4	Assignment in Fig. 4
– The process when a non-positive integer is entered – Number of loops – Forgot the % specifier – Missing "&" in the scanf function – Missing the carriage return – Triangle could not be stuffed right – Calling a function in a double loop	– Forget to use float type – No idea to branch tax categories and point additions with flag variables – Not aware of using the structure – Not aware that the array of structure – No idea how to use the structure – Not aware of using the array – Enumerate the goods without loops – Wrong specifier for floating point – How to cast / rounding off

that the learner's incomprehensible factors can be investigated. The explanatory variables used in the proposed method are roughly divided into three categories:

- Attention to each block
  - The time at which each block is touched in the first time
  - The time spent for watching each block description by hover
  - The number of times each block has been dragged and dropped
  - The frequency of drag-and-drop of each block in the whole answering time
  - The frequency of drag-and-drop of each block in each quarter of the answering time
  - The dispersion of the time watching the description
  - the dispersion of the number of drag-and-drop times of each block
- Correctness in placing blocks
  - Correctness of input to each blank
  - The number of times each blank correction failed
  - How close the organization of whole blocks is to the model answer in the editing distance
  - The number of times overall blank correction failed
- The time spent for answer
  - The total answering time
  - The ratio of the drawing time on the screen display toward the time watching assignment sentence screen
  - Number of times to switch between the assignment sentence screen and the drawing screen

The attention level and correctness of each block reveal the block that confuses learners. The overall correctness reveals how accurately the blocks are combined. In addition, the time and the ratio spent on the drawing using PAD indicate how much learners got confused during the drawing.

4.4 Objective and Method of Experiment

The purpose of this experiment is to clarify the understanding considering the learner's programming thinking ability from the learner's operation history using code puzzles. The subjects are 17 university students, including first-year undergraduate students who started learning programming and first-year graduate students who are accustomed to programming. Each of the subjects solved a given programming task. The time to solve the task took the minimum of 7 min and the maximum of 40 min. The degrees of attainment were various between the subjects. After working, for the labeling described later, we conducted a questionnaire asking about the cognitive load of this task and usual experience. While solving the task, the instructor did not give any hints and just watched to measure programming thinking ability.

4.5 Actual Understanding and Cognitive Load Measurement

Before performing analysis by machine learning, the instructor labeled the objective variables used in supervised learning. The label used for training is a binary value (0/1) indicating whether or not the subject understood. For subject, a 0 or 1 label was assigned to each subject for each of the learning item achievement and programming thinking ability.

1. The assignment content was very complicated.
2. The knowledge on filling in the blanks and functions was very complicated.
3. The concepts and ideas in the task were very complicated.
4. It was very unclear how to use tools and PAD notation.
5. It was difficult to understand how to use tools and PAD notation.
6. Tools and PAD notation are very inefficient from a learning perspective.
7. The task has improved my understanding of programming.
8. The task improved the understanding of programming process
9. The task has increased my understanding of programming concepts and definitions.
10. The task has improved my knowledge of programming.
11. It took a lot of mental effort because the task was complicated.
12. Due to the explanation in the task and the usage of the tool, I took a lot of mental effort.
13. It took a lot of effort to improve knowledge and understanding in the task.
14. Relative programming score
15. Relative programming experience

Fig. 6. Questionnaire for measuring cognitive load

The questionnaire answer in the experiment was used for labeling. Figure 6 shows the list of used questionnaires. It includes the items to investigate the cognitive load [16] felt by the subject while solving the task. This questionnaire is a question group based on a 10-point Likert scale. The subject answers the question sentence subjectively. Generally, if the learner does not misunderstand a task, a high cognitive load means a low understanding of the task. However, in reality, there were many contradictory in the answer to the questionnaire and the performance of the task. In other words, there were many subjects who did not perform the task well despite the small cognitive load. This suggests that some subjects misunderstood the content of task. From this, it can be said that it is difficult to evaluate understanding, especially programming thinking ability, unless the instructor watches the learner's behavior of the process of solving the task. Therefore, in the labeling, with referring to the questionnaire, the instructor give the labeling from the viewpoint of both the learning item achievement and the programming thinking based on the behavior during task.

4.6 Estimation of Understanding and Distribution of Learners

Based on the operation history data collected from the 17 subjects and the labels given by the instructor, the learning item achievement and programming thinking ability were classified. Both the random forest and logistic regression are used for classification. With the random forest, high classification accuracy and variable importance can be obtained. On the other hand, with the logistic regression, variable importance based on regression coefficient and understanding by values from 0.0 to 1.0 can be obtained. For the analysis, we used Scikit-learn in Python to find the optimum parameters by the grid search and ensured the generalization performance by using the Leave-one-out cross-validation.

Tables 2 and 3 show the accuracy rates of the model created by this experiment. The accuracy rates of both models exceeds 80% and it can be said that the models have a high degree of accuracy.

Table 2. Performance of learning item achievement classification models

Logistic regression		Random forest	
Accuracy	0.82	Accuracy	0.82
Precision	0.78	Precision	0.86
Recall	0.88	Recall	0.75
F1-score	0.82	F1-score	0.80

Table 3. Performance of programming thinking classification model

Logistic regression		Random forest	
Accuracy	0.82	Accuracy	0.94
Precision	0.79	Precision	0.92
Recall	1.00	Recall	1.00
F1-score	0.88	F1-score	0.96

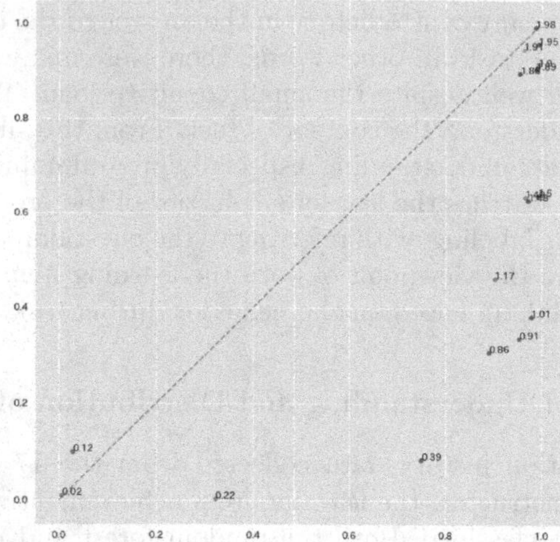

Fig. 7. learning item achievement and programming thinking ability

Additionally, the distribution of the subjects' understanding is displayed on the two-dimensional coordinates using the two axes of the classification probabilities of learning item achievement and programming thinking ability predicted by the logistic regression. The distribution map is shown in Fig. 7. The label of the distribution map is the degree of schema construction of the corresponding subject. In the distribution map, the subjects close to the diagonal line of 45 degree show that the degree of learning items achievement and the programming thinking ability are similar and that the two abilities are well balanced. Among them, the subjects located in the upper right shows that both abilities are high. On the other hand, learners distant from the 45 degree line to the lower side have a high learning item achievement but have a low programming thinking ability, which indicates a poor balance. If the schema construction degree is used, those who have high knowledge are evaluated only to some extent and those who have a balance of both values are highly evaluated.

5 Consideration

5.1 Correlation Between Schema Build and Grade

The validity of the schema construction level is confirmed by comparing the calculated schema construction level with the results of the actual class performance. Table 4 shows the calculated schema construction level of the subjects and the total scores of the 32 fill-in-the-blank tasks that the subjects have took in the actual classes. The correlation coefficient among them is also shown. The subjects were those who were able to obtain actual class performance data among the 17 subjects. In addition, we excluded those that could not be tested due

Table 4. Correlation with the calculated schema construction level and a list of grades

Subject	A	B	C	D	E	F	G	H	I	Correlation
Schema level	1.922	1.887	1.885	1.510	1.502	1.016	1.001	0.430	0.215	
Score	754	760	736	713	769	723	775	542	674	0.67

Table 5. Top 10 important variables by logistic regression and random forest

[1]List of important variables for classifying learning item achievement

	Regression coefficient	Explanatory variable	Block detail
1	0.64	touch14	1 character printing
2	0.43	Blank 9 Similarity	Number of printing
3	0.42	touch19	Line break
4	0.35	hover18	For loop
5	0.34	Blank 1 Similarity	Num of loop specifying
6	-0.35	touch18	For loop
7	-0.37	Blank 3 Similarity decrease	Variable specifying
8	-0.40	Tab switching	
9	-0.42	touch12	Variable defining
10	-0.55	first12	Variable defining

	Variable importance	Explanatory variable	Block detail
1	0.15	touch14	1 character printing
2	0.14	first17	For loop(Trap)
3	0.10	Touch frequency ~ 1/4 period	
4	0.09	Touch frequency ~ 3/4 period	
5	0.09	hover17	For loop(Trap)
6	0.08	touch19	Line break
7	0.07	first13	For loop
8	0.06	hover20	line number printing
9	0.05	Touch frequency	
10	0.04	first8	console printing

[2]List of important variables for classifying programming thinking ability

	Regression coefficient	Explanatory variable	Block detail
1	0.73	Model answer similarity	
2	0.45	Blank 11 Similarity	Argument specifying
3	0.39	Blank 5 Similarity	Specifier specifying
4	0.39	first10	Printing(Trap)
5	0.35	first5	Variable defining
6	0.33	Blank 6 Similarity	Line break
7	0.32	Blank 12 Similarity	Condition specifying
8	-0.37	Blank 9 Similarity decrease	Number of printing
9	-0.47	Touch frequency distribution	
10	-0.72	hover14	1 character printing

	Variable importance	Explanatory variable	Block detail
1	0.18	first5	Variable defining
2	0.17	touch13	For loop
3	0.12	hover14	1 character printing
4	0.12	hover12	Variable defining
5	0.12	first17	For loop(Trap)
6	0.10	touch11	Variable defining(Trap)
7	0.06	hover15	Function defining
8	0.05	Percentage of Drawing time	
9	0.05	touch16	Character input
10	0.03	Blank 8 Similarity decrease	Number of printing

touch: Number of drag and drop
hover: Time spent looking at the code block description
first: How fast the code block was first adopted

to poor physical condition and excluded weeks of tasks that were not directly related to programming.

The correlation coefficient of 0.67 is not very high. However, it correlates to some extent with the credible data of scores. This shows the validity of the degree of schema construction. The reason why a high correlation does not appear is that the data of the performance of the fill-in-the-blank tasks does not consider programming thinking ability. On the other hand, the proposed method easily quantifies the real programming ability considering programming thinking ability.

5.2 Consideration from Important Variables

We compare the classification models of learning item achievement and programming thinking ability and consider the differences. Table 5 shows the important variables of the models created in this experiment.

The subjects with low learning item achievements moved many variable blocks and loop blocks by focusing on the variables related to "touch" and "hover". From this, it is said that they did not understand how to use variables and could not answer the basic parts such as loop statements. It is considered that this is because the subject with low learning item achievement could not think the meaning of the variable name. In addition, the learning item achievement is lower when the iterator variable "i" is declared earlier and the touch frequency up to 1/4 h is higher. This suggests that the learners who worked on the task sooner after provision of the task got confused. It is considered that they have repeatedly switched between the drawing screen and the task display screen. On the other hand, the subjects with high programming thinking ability took less time to adopt the "input block". From this, it is considered they immediately grasped the meaning from the name of the variable and adopt it. Moreover, since they read the explanation of the variable especially "loop block" carefully, it can be said that they tend to think carefully how many times to loop.

In the learning item achievement level, the "one-character printing block", which is the deepest part of the loop, is important. From this, it can be seen that the subjects with low learning item achievement could not reach to think this module. On the other hand, in programming thinking, the time spent for watching the explanation of single-character printing is important. This suggests that the subjects with high programming thinking have reached the deepest part of the loop and considered it. In addition, the variables related to the touch frequency is important in the learning item achievement. This indicates that the subjects with low learning item achievement tended to assemble modules without thinking. In programming thinking ability, the variable of touch frequency is important. In other words, it was found that the subjects without programming thinking touched many modules that were not necessary for the task. They have been at a loss. In addition, there are many important variables that indicate "looking at the module description". This shows that the subjects with high programming thinking ability firmly identified the modules to be used at first and established the course before tackling the task.

The important variables for the programming thinking ability can indicate similarity between the model and the subject's answers. This shows that an ideal answer cannot be achieved only with the learning item attainment level. It is because programming thinking is indispensable for solving code puzzles [12]. In addition, the variable of "the time of looking at the explanation of 1-character printing" is important. This results implies that the subjects who were confused about the matters such as "1-character printing" may be greatly evaluated negatively. Moreover, since "designation of arguments and conditions" is important, it can be said that a learner with high programming thinking ability grasps the flow of data and chooses appropriate variables.

The followings are the findings for learners of low understanding for programming. Learners with a low level of understanding cannot interpret how to use variables and loops which are basic matters related to the degree of achievement of learning items. In addition, there is a tendency to tackle the task immediately without careful interpretation of assignment sentenses. The ablility of the programming thinking is evaluated low when the answer is far from the model answer. It gets lower if learners are stumled at basic matters. Learners poor at the programming thinking fail to look deeply at the explanation. They are likely to touch many blocks because they have no perspectives to their answers. Furthermore, they cannot construct structures suitable for right data flow, which prevents them from specifying arguments and conditions properly. They do not look at the explanations deeply, touch many blocks, and do not set the course for answers. In other words, they cannot think deeply. Furthermore, they cannot assemble the structure firmly. They do not understand the data flow and cannot specify arguments or conditions. From these facts, it is considered that the learning item achievement is the basic ability while the programming thinking ability is the comprehensive ability including the learning item achievement.

5.3 Usefulness in Educational Settings

The results of this experiment show that it is possible to estimate the comprehension level considering the programming thinking ability from the learner's operation history when answering the code puzzle. At present, there are many one-to-many forms in educational settings such as universities or companies. Though they sometimes hire multiple assistants to support learners, the current situation is that the number of assistant is insufficient. Therefore, it is difficult for the instructor to grasp the understanding level of all learners.

The schema construction estimated by the proposed method can measure the understanding of many learners at once if a model is learned by labeling dozens of samples. In particular, the classification probability output by the logistic regression is a value from 0.0 to 1.0. Therefore, by setting a threshold appropriately, we can discriminate between learners who can be able to solve the task and learners who need guidance.

In the field of education, it is common to learn various elements with multiple tasks. The schema construction calculated for multiple tasks over a period can be visualized by studying the transitions of learning progress and growth. Moreover, by referring to the important variables, the instructor can consider where the learner stumbled. For example, in the task of this experiment, it can be seen that many learners did not understand what the meaning of the variable has from its name because the importance of the block for the variable was high. Since the importance of variables on blocks regarding the number of loops was high, it can be seen that many learners could not fully consider the processing flow.

6 Conclusion

6.1 Research Contribution

This paper proposed a method to estimate the learner's programming thinking ability as well as learning item achievement by analyzing his/her answering process of the code puzzles.

In order to measure programming ability precisely, it is necessary to provide the learners an environment like a code puzzles where he/she can focus on combining programming elements and to extract his/her answering process and behavior. The estimation is performed by learning the random forest and logistic regression models, where the objective variables are programming thinking or learning item achievement levels, respectively, and the explanatory variables are learner's actions in solving code puzzles.

This paper's question is: How effective is the method that can estimate the learner's understanding situation by considering programming thinking ability.

As the result of the experiment, it was found that the proposed method was able to estimate the understanding with the accuracies of more than 80%. In addition, considering the difference between the learning item achievement and the programing thinking ability from the difference of the variable importance, it was confirmed that the programming thinking ability is based on the learning item achievement and that the programming ability cannot be measured only by the learning item achievement. Furthermore, based on the schema theory [13], we defined the schema construction level from two labels, learning item achievement and programming thinking ability.

The results of this experiment show that the learner's programming ability can be measured more accurately by considering the learner's logical constructive ability in the code puzzle rearrangement assignment. The accurate measurement of the learner's programming ability contributes to developing the learner's true programming ability, which cannot measured by only the score of tests. In addition, the importance of each variable in the behavior analysis leads to the identification of learner's misunderstanding factors and the improvement of class contents.

6.2 Issues and Future

Labeling is not a small cost in the educational setting. However, this is unavoidable as long as supervised learning is used. Blikstein *et al.* [17] suggested that the understanding of learners can be measured by using the clustering method with programming structure. Therefore, in our research, there is a possibility that the understanding and the factor of misunderstanding can be estimated by labeling using such the clustering method. The model created in one year can be used in the following years as long as the tasks are the same and the level of learners is the same. In general, at one university, the levels of students are almost the same for several years. Furthermore, the distribution of the programming ability of students as described in this paper can be visualized based on the learning item achievement and programming thinking ability calculated by the model. Furthermore, there is a program that visualizes the students' behavior. With visualization tools, it is not difficult to label for multiple people. This also allows refinement of the model.

The paper shows that the analysis method based on code puzzles can estimate the comprehension level that takes into account programming thinking. Our next goal is to enhance contents of feedbacks. In this method, we analyzed the comprehension level from statistical data on specific time points. Therefore, the content of the feedback is also based on statistical data on specific time points. However, programming thinking relates to the construction of several pieces of knowledge. Program codes are gradually constructed in the process of solving assignments. The codes in specific time points are built to contribute to codes in their succeeding time points. There should be temporal dependency from learner behavior to build the former to that to build the latter. That is, when learners with programming thinking solve programming assignments, they always have perspectives of what they are going to do. This perspective will change as the progress of the code building. Its transition can be represented with a state transition diagram. State transition diagrams have already been successfully introduced into education in works such as knowledge-tracing. In this study, some code puzzles have turned out to work well to estimate the comprehension level of learners. As the next work, state transition diagrams are introduced to analyze learner behavior solving them. Letting the shape of codes and learner behavior to build it are observable data, the perspective of learners can be regarded as their internal states in the state transition model. Learners would fall into trouble when they lose the right perspectives. The analysis would enable us to visualize the development process of perspectives of individuals according to their behavior. It facilitates us to clearly identify which parts of the process make them lose their perspectives.

References

1. Hellas, A., et al.: Predicting academic performance: a systematic literature review. In: Proceedings of Companion of the 23rd Annual ACM Conference on Innovation and Technology in Computer Science Education ITiCSE, pp. 175–199 (2018). https://doi.org/10.1145/3293881.3295783
2. Whipkey, K.L.: Identifying predictors of programming skill. SIGCSE Bull. **16**(4), 36–42 (1984). https://doi.org/10.1145/382200.382544
3. Mazlack, L.J.: Identifying potential to acquire programming skill. Commun. ACM **23**(1), 14–17 (1980). https://doi.org/10.1145/358808.358811
4. Nesbit, J.C., Liu, L., Liu, Q., Adesope, O.O.: Work in progress: intelligent tutoring systems in computer science and software engineering education. In: ASEE Annual Conference & Exposition, pp. 1–12 (2015)
5. Ministry of education. Elementary programming education guide (second edition). https://www.mext.go.jp/component/a_menu/education/micro_detail/icsFiles/afieldfile/2018/11/06/1403162_02_1.pdf
6. Crow, T., Luxton-Reilly, A., Wuensche, B.: Intelligent tutoring systems for programming education: a systematic review. In: Proceedings of the 20th Australasian Computing Education Conference, pp. 53–62 (2018). https://doi.org/10.1145/3160489.3160492
7. Jadud, M.C.: Methods and tools for exploring novice compilation behaviour. In: Proceedings of the Second International Workshop on Computing Education Research, pp. 73–84 (2006). https://doi.org/10.1145/1151588.1151600

8. Mysore, A., Guo, P.J.: Porta: profiling software tutorials using operating-system-wide activity tracing. In: Proceedings of the 31st Annual ACM Symposium on User Interface Software and Technology, pp. 201–212 (2018). https://doi.org/10.1145/3242587.3242633
9. Guo, P.J.: Codeopticon: real-time, one-to-many human tutoring for computer programming. In: Proceedings of the 28th Annual ACM Symposium on User Interface Software & Technology, pp. 599–608 (2015). https://doi.org/10.1145/2807442.2807469
10. Asai, S., Dong Phuong, D.T., Shimakawa, H.: Identification of factors affecting cognitive load in programming learning with decision tree. J. Comput. **14**(11), 624–633 (2019). https://doi.org/10.17706/jcp.14.11.624-633
11. Ihantola, P., Sorva, J., Vihavainen, A.: Automatically detectable indicators of programming assignment difficulty. In: Proceedings of the 15th Annual Conference on Information Technology Education, pp. 33–38 (2014). https://doi.org/10.1145/2656450.2656476
12. Parsons, D., Haden, P.: Parson's programming puzzles: a fun and effective learning tool for first programming courses. In: Proceedings of the 8th Australasian Conference on Computing Education, vol. 52, pp. 157–163 (2006)
13. Schnotz, W., Kürschner, C.: A reconsideration of cognitive load theory. Educ. Psychol. Revi. **19**, 469–508 (2007). https://doi.org/10.1007/s10648-007-9053-4
14. Futamura, Y., Kawai, T., Horikoshi, H., Tsutsumi, M.: Development of computer programs by problem analysis diagram(pad). In: Proceedings of the 5th International Conference on Software Engineering, pp. 325–332 (1981). https://doi.org/10.5555/800078.802545
15. Coffey, J.W.: Relationship between design and programming skills in an advanced computer programming class. J. Comput. Sci. Coll. **30**(5), 39–45 (2015)
16. Morrison, B.B., Dorn, B., Guzdial, M.: Measuring cognitive load in introductory CS: adaptation of an instrument. In: Proceedings of the Tenth Annual Conference on International Computing Education Research, pp. 131–138 (2014). https://doi.org/10.1145/2632320.2632348
17. Blikstein, P., Worsley, M., Piech, C., Sahami, M., Cooper, S., Koller, D.: Programming pluralism: using learning analytics to detect patterns in the learning of computer programming. J. Learn. Sci. **23**(4), 561–599 (2014). https://doi.org/10.1080/10508406.2014.954750

Numerical Methods of Solving Management Problems

Wind Farms Maintenance Optimization Using a Pickup and Delivery VRP Algorithm

Vincenza Carchiolo[1(✉)], Alessandro Longheu[2], Michele Malgeri[2], Giuseppe Mangioni[2], and Natalia Trapani[2]

[1] Dipartimento di Matematica ed Informatica, Universitá degli Studi di Catania, Catania, Italy
vincenza.carchiolo@unict.it

[2] Dipartimento Ingegneria Elettrica Elettronica Informatica, Universitá degli Studi di Catania, Catania, Italy
{alessandro.longheu,michele.malgeri,giuseppe.mangioni, natalia.trapani}@unict.it

Abstract. Operations and maintenance of wind farms in renewable energy production are crucial to guarantee high availability and reduced downtime, saving at the same time the cost of energy produced. While SCADA or NLP-based techniques can be used to address maintenance tasks, efficient management of wind farms can be really achieved by adopting an intelligent scheduling algorithm. In this paper an algorithm that optimizes maintenance intervention routing is presented, taking into account the location of spare parts inventory, geographically dispersed intervention sites, and overall costs of the intervention, considering human resources and fuel consumption. Different scenarios are discussed through a toy example, to better explain the algorithm structure, and a real case of wind farms distributed in Sicily, to validate it. The usefulness of the proposed algorithm is shown also through some Key Performance Indicators selected from UNI EN 15341:2019. The purpose of this work is to show the effectiveness of adopting a VRP algorithm in optimizing the maintenance process of wind farms by investigating real scenarios; in addition, the proposed approach is also efficent therefore feasible for coping with unplanned interventions changes.

Keywords: Maintenance · KPI · Optimization · Renewable energy · Power plants · Vehicle routing problem

1 Introduction

Wind farms for renewable energy production are complex systems that requires an effective and efficient maintenance to fulfill their goal while saving the overall cost of the energy generated [1,2].

E. Ziemba and W. Chmielarz (Eds.): ISM 2020/FedCSIS-IST 2020, LNBIP 413, pp. 67–86, 2021.
https://doi.org/10.1007/978-3-030-71846-6_4

Operations and Maintenance (O&M) of a wind farm are often outsourced by the owner to a provider [3] that has to grant high availability of wind turbines, reducing downtime by optimizing maintenance scheduling. Then an O&M provider manages a multi-location maintenance service which requires to ensure preventive maintenance, to perform corrective maintenance as soon as possible, to detect performance degradation or incipient failures through condition-based and predictive maintenance, and to take advantage of opportunities that may arise in the context of a wind farm when a scheduled downtime or a failure of a system close to the item of interest occurs [4]. In fact, a wind turbine consists of approx 15–20 000 components and many of them affect each other even if they are not directly connected.

Remote monitoring through a Supervisory Control and Data Acquisition System (SCADA) gathers information about the behavior of the single wind turbine as well as the whole wind farm, providing components performance data [5,6] and error signals [7–9].

Other information on turbine reliability can be extracted by maintenance reports compiled by operators. Using techniques based on Natural Language Processing (NLP) meaningful information can be inferred from the semi-structured text and raw notes provided by maintenance operators, for instance in [10] monitoring and historical data were used to predict the failures in order both to plan the maintenance team missions as well as the need of spare parts.

Long term operating strategies like the creation of large inventories or increasing the dedicated response time can be not enough to gain a competitive advantage; rather the efficient maintenance management of wind farms can be achieved only if a distributed spare parts storage is supported by an intelligent scheduling algorithm that permits to reduce costs and downtime, therefore leading to company's success [11].

This work starts from results of the WEAMS project [12,13] and deals with the development of an innovative asset management platform for the wind industry to analyze maintenance management performance. The project addressed some logistic related questions, e.g. different strategies to limit costs and downtime due to preventive and corrective maintenance. In particular in this paper we present the algorithm that optimizes maintenance intervention routing taking into account the location of spare parts inventory, geographically dispersed intervention sites (wind farms), and overall costs of the intervention based on human resources and fuel consumption. To show the usefulness of the algorithm in generating better maintenance management performance some Key Performance Indicators selected from UNI EN 15341:2019 have been analysed, comparing scenarios with and without algorithm results.

The paper is organized as follows. In Sect. 2 the problem to be addressed is outlined and compared with other works existing in literature. In Sect. 3 the pickup-and-delivery algorithm with time windows used to manage wind turbine spare parts delivery is described in detail. Section 4 presents a couple of experiments in different scenarios, the former being a toy example to show how the algorithm works whereas the latter refers to a real case scenario with power plants located in Sicily. Finally, Sect. 5 briefly summarizes our conclusions, open questions and future directions.

2 Literature Review

2.1 Maintenance and Logistics in Wind Farms

Maintenance in wind farms presents some typical matters that were studied from different point of view. First of all, both offshore and onshore wind farms are characterized by the difficulty of being reached: the former due to the harsh marine environment, the latter due to the orography typical of the land where they are installed, e.g. on the top of mountains or hills in isolated areas. Maintenance interventions can be scheduled in a specific time window according to a set of factors such as energy demand, electricity market price, and wind speed [14,15]. Moreover, the planned maintenance inspections or interventions can be limited by external events such as heavy rain, snow and/or wave motion in the case of offshore wind farms.

Maintenance service planning needs to know the set of preventive and opportunistic maintenance (PM) actions, maintenance interventions' duration, and required resources (including supplies, workforce, and spare parts) for multi-located production sites [4,16,17].

Whenever several wind farms must be managed, especially if they are geographically distributed on a large scale, also in onshore installation it becomes necessary to pay attention to elements related to the logistic service (correct spare parts and maintenance materials, timing, routing efficiency), taking into account that preventive maintenance with online condition monitoring is far more efficient in a life cycle view than a corrective maintenance policy [18]. In fact, each wind turbine is a complex multi-component system where a single component failure or performance reduction allows to implement preventive opportunistic maintenance for the others, thus reducing the losses of accidental failures [19,20].

To make it possible, constraints deriving from both maintenance interventions scheduling (time and resources) and spare parts storage and delivery should be considered [21,22] in order to limit system downtime and contain overall maintenance costs. While the former issue can be tackled in traditional ways, the latter is quite complex due to the large geographical distribution of plants, often hard to reach. Moreover, some spare parts are very large and heavy objects, difficult to move from storage to plant even if their substitution is rare; other components, instead, exhibit very low capital costs but they are expensive throughout the life cycle because their substitution (e.g. bearings, sensors, electrical cables) or consumption (e.g. lubricants) is frequent during preventive maintenance also because their failure can generate significant power production losses.

Among other typical problems of power plants maintenance, wind farms managers must also tackle the travel times of the workers needed to reach the site and the transport of spare parts in a location often far and difficult to reach. Moreover, the maintenance technicians of a wind farm are often highly specialized and they focus only on specific parts of the product, thus generating high operational expenditures (OPEX). Therefore, to stock a lot of large and heavy spare parts in several places could result in high capital costs (CAPEX). An effective maintenance strategy must take into account not only multiple stakeholders for production processes and locations but also the movements of spare parts and workforce (transportation costs) and reduction of the indirect cost of parts stored in a warehouse.

The classical optimization of maintenance for wind farms spans over six main categories: facility location and demand allocation [23], warehouse and spare parts management [24], equipment supply chain [25], logistic networks complexity analysis and network performance measurement [26], scheduling and maintenance routing optimization. Some study proposes an approach for thoroughly investigating the effects of various factors such as government subsidies, lost power generation, and location on maintenance cost using a mathematical deterioration model of offshore wind turbine [27].

In the case presented in this paper we consider the optimization of the logistic network of green energy production, like wind and solar plants, aiming at reducing the overall cost (i.e. number of vehicles and maintenance team, fuel consumption, etc.) while complying with the time constraints.

According to Zhu et al. [28] maintenance planning must satisfy several objectives such as system availability performance reachable with maintenance resources constraints, regulatory requirements, and opportunities (in the short-term horizon) of favorable weather conditions.

The problem we address is twofold: modeling the network using complex network theory, and optimizing costs while respecting constraints through the use of Vehicle Routing Problems (VRP) optimization techniques. Related works on maintenance scheduling optimization algorithm are discussed in Sect. 2.2.

Maintenance management performance can be measured through a set of Key Performance Indicators (KPIs), that are meaningful quantitative or qualitative measure of systems or processes features having an analytical formulation. The standard UNI EN 15341:2019 [29] designed a framework of influencing factors such as economic, technical and organizational aspects, to evaluate and improve maintenance efficiency and effectiveness to achieve excellence. A set of KPIs allows to measure the state of a system, making comparisons (internal and external benchmarks), make a diagnosis of maintenance management system, identify the objectives and define the goals to be achieved, plan improvement actions, continuously measure changes over time. In the context of wind farms is appropriate to define an organizational model of the maintenance function in relation to the required objectives, available resources and existing constraints. In the context of this paper just some KPIs can be considered meaningful, i.e. some of those related to maintenance for asset management (encoded as PHA in UNI EN 15341:2019), maintenance management subfunction (encoded as M), personnel competencies subfunction (encoded as P), maintenance engineering subfunction (encoded as E), organization and support subfunction (encoded as O&S), administration and procurement subfunction (encoded as A&S). The definitions and assessment of some of them are shown in paragraph 4 to support the evaluation of the usefulness of the proposed algorithm.

2.2 Maintenance Scheduling Optimization

A lot of recent scientific literature is available on maintenance scheduling for offshore wind farms trying to solve both maintenance and logistics matters. The work [21] provides a conceptual classification framework for the available literature about maintenance strategy optimization and inspection planning of wind energy systems. The effort typical of maintenance tasks in wind farms is due to several factors, as both space-related constraints, e.g. the difficulty of reaching off-shore (but also many on-shore) locations as well as time-related constraints, e.g. when trying to accomplish maintenance on the "right" day (as indicated by optimization algorithms) but a wind storm just hit that area.

Some authors studied the preventive maintenance scheduling problem of wind farms in the offshore wind energy sector which operates under uncertainty due to the state of the ocean and market demand formulating a fuzzy multi-objective non-linear chance-constrained programming model to obtain a trade-off between reliability maximization and cost minimization objectives [30].

An additional related matter is the logistics of spare parts, whose management significantly affects maintenance tasks; having the right part at the right time in the right location is critical to guarantee business continuity and maintenance performance. In [31], authors focus mainly on weather conditions to determine the best time window and execution order for optimal intervention. Similarly, [32] proposes a hybrid heuristic optimization of maintenance routing and scheduling in particular for offshore wind farms, where optimal vessel allocation scheme is crucial (though spare parts are not considered). In [33], offshore wind farms are also addressed, in this case finding the best routes for the crew transfer vessels and in [34] for vessel fleet composition. Other authors propose an integrated framework for condition monitoring, diagnosis, and maintenance of wind turbines in which a module is able to perform the detailed scheduling of maintenance operations at a set of wind farms maintained by common personnel, through a mixed-integer linear program [35]. The system constitutes a module of an integrated framework for condition monitoring, diagnosis, and maintenance of wind turbines Conversely, the work [36] focuses on on-shore wind farms and considers forecast wind-speed values, multiple task execution modes, and daily restrictions on the routes of the technicians to determine optimal maintenance operations scheduling.

One of the most investigated combinatorial optimization problem concerns vehicle routing (VRP) [37], we adopted here for our purpose. It consists of finding a route sets so that the vehicles can optimally serve customers' requests (according to a specific function to be optimized) while respecting constraints. The interest in solving VRP problems is motivated by their practical relevance and their inherent difficulty. Of course, the difficulties grow up in the presence of great distance and hard to reach places. To overcome the problem complexity various heuristics have been developed to produce good solutions with tractable computational complexity [32,33,38]. The problem indeed presents significant computational challenges by admitting, in its more general formulation, further constraints such as the respect of time windows on both customers and deposits or by imposing a maximum vehicle transport load capacity and a maximum speed.

All these works tackle the question with different approaches, for instance [33] is based on the Large Neighbourhood Search meta-heuristic, whereas [36] adopts linear programming formulations and branch-and-check approach. In [31] the optimization is achieved simply through brute force whereas [32] adopts a hybrid optimization using first mixed particle swarm optimization to determine an optimal vessel allocation scheme and then discrete wolf pack search (DWPS) to optimize the maintenance route according to all constraints. A common feature most works share is the exploitation of real historical datasets to achieve realistic optimizations.

3 Pickup and Delivery VRP with Time Windows

As introduced previously, in this work we address the maintenance plan optimization problem by mapping it on a specific VRP problem. In detail, we employ a pickup and delivery VRP with a time windows algorithm to take into account all the constraints imposed by our specific problem. It is known that determining the optimal solution to VRP is NP–hard, hence to approach such a problem many heuristics have been developed. Here we employ the algorithm proposed in [39], which consists of two phases. It is indeed recognized that in a typical VRP minimizing the objective function in a single phase might not be the most efficient way to decrease the number of routes and vehicles because such function leads many times to solutions with low travel costs and this could make it difficult to reach solutions with few routes but with a higher travel cost.

To avoid this, the above-mentioned algorithm uses a two-stage approach consisting in

1. the minimization of the number of routes through the use of a Simulated Annealing algorithm.
2. the minimization of the total travel cost by using a Large Neighborhood Search algorithm.

In the following, we present the Pickup and Delivery Vehicle routing problem with time windows (PDPTW) by first introducing some definitions taken from [39].

Customers. The problem is defined in terms of the N customers, represented by the numbers $1, ..., N$, and a deposit, represented by the number 0. In general, with the term site, we identify the N customers and the deposit as well, i.e. sites range from 0 to N.

- $Customers^p$ denotes the set of withdrawal points (pickup customers). Specifically, in our application these are warehouses where are stored materials and maintenance spare parts dedicated to wind energy power plants located in a certain area.

- $Customers^d$ indicates the delivery points (delivery customers). In our simulation scenario, these are the plants (wind farms) requiring maintenance interventions.

Travel Cost. The cost of the path between the generic sites i and j is indicated with c_{ij}. It is supposed that such a cost must satisfy the triangular inequality: $c_{ij} + c_{jk} >= c_{ik}$. The normalized travel cost c'_{ij} is also defined as the cost c_{ij} between sites i and j divided by the max cost among all couple of sites. The travel costs are associated with the time for maintenance workforce, and with the distance for fuel consumption. In fact, the difficulties in reaching the sites not only depends on where they are located but it is also affected by the efficiency of road infrastructure, therefore we preferred not to merely rely on the distance rather we considered the estimated actual time needed to reach each place.

Service Time. A service time is also associated with every customer i, together with a demand $q_i \geq 0$. If i is a pickup customer, the delivery counterpart is denoted by $@i$. Given that, the demand of $@i$ is $q_{@i} = -q_i$.

Vehicles. In this problem, we suppose to have m identical vehicles each having capacity Q.

Routes. In general, a route starts from the depot, visits a certain number of customers at most once, and finally returns to the depot, i.e. a route is a sequence $\{0, v_1, ...v_n, \}$, where v_i is the generic vertex of the path. Note that in a route all v_i are different, i.e. each vertex is touched only once (excluding the depot). Given a route $r = \{v_1, ...v_n, \}$, we denote with $cust(r)$ the set of its customers, i.e. $cust(r) = \{v_1, ...v_n, \}$, With $route(c)$ we denote the route the customer c belongs to. For a given route r, its length is indicated by $|r|$, while the number of visited customers is denoted by $|cust(r)|$. The travel cost of a route is indicated by $t(r)$ and represents the cost of visiting all of its customers; it is defined as:

$$\begin{cases} t(r) = c_{0v_1} + c_{v_1v_2} + ... + c_{v_{(n-1)}v_n} + c_{v_n0} & \text{if route !=}\emptyset \\ 0 & \text{otherwise} \end{cases} \tag{1}$$

Routing Plan. It is a set of routes $\{r_1, ..., r_m\}$ with $(m \geq N)$ visiting all customers exactly once:

$$\begin{cases} \bigcup_{i=1}^{m} cust(r_i) = Customers \\ cust(r_i) \cap cust(r_j) = \emptyset \qquad (1 \leq i < j \leq m) \end{cases} \tag{2}$$

A routing plan assigns a single successor and predecessor to every customer. Given a routing plan σ and a customer i, $succ(i, \sigma)$ and $pred(i, \sigma)$ are respectively the predecessor and the successor of i in the routing plan σ (shortly indicate as i^+ and i^- in the following).

Time Windows. Each site is associated with a temporal window $\{e_i, l_i\}$, where e_i represents the earliest arrival time and l_i the latest arrival time. This means that a vehicle can arrive on a site i before e_i, but it must wait for e_i to start the service. Vehicles must arrive at any site i before the end of the time window l_i. In the specific case of the depot, its temporal window $[e_0, l_0]$ individuates the time e_0 in which all vehicles leave the depot and the time l_0 when all vehicles return to the depot. The departure time δ_i of a given customer i is defined as:

$$\begin{cases} \delta_0 = 0 \\ \delta_i = max(\delta_{i^-} + c_{i^-i}, e_i) + s_i \quad (i \in Customers) \end{cases} \tag{3}$$

The Earliest Service Time a_i of a given customer i is defined as:

$$a_i = max(\delta_{i^-} + c_{i^-i}, e_i) \quad (i \in Customers) \tag{4}$$

The Earliest Arrival Time $a(r)$ of a route r is defined as:

$$a(r) = \begin{cases} \delta_{v_n} + c_{v_n 0} & if\ (route! = \emptyset) \\ e_0 & otherwise \end{cases} \tag{5}$$

For a customer i the time window constraint is satisfied if $a_i \leq l_i$ and, in particular the time window constraint for the deposit is satisfied if $a(r) \leq l_0\ \forall r \in \sigma$.

Capacities. Let us define the demand of a route r at customer c as:

$$q(c) = \sum_{i \in cust(r)\ \&\ \delta_i \leq \delta_c} q_i \tag{6}$$

With the constraint that for a customer c, $q(c) \leq Q$.

PDPTW. A solution to the PDPTW is a routing plan σ that satisfies all the following constraints:

$$\begin{cases} q(i) \leq Q \\ a(r_j) \leq l_0 \\ a_i \leq l_i \\ route(i) = route(@i) \\ \delta_i \leq \delta_{@i} \end{cases} \tag{7}$$

where $i \in Customers$ and $1 \leq j \leq m$.

A solution to the PDPTW consists in finding a routing plan σ satisfying the above-mentioned constraints that also minimizes the number of vehicles and, in case of ties, the total travel cost. Formally, σ minimizes the following objective function:

$$f(\sigma) = \langle |\sigma|, \sum_{r \in \sigma} t(r) \rangle \tag{8}$$

The algorithm used to find a solution to the PDPTW is that proposed in [39]. Its first stage performs the minimization of the number of routes via a simulated annealing algorithm, i.e. it starts from a solution and then produces a new random solution that is accepted with a probability that depends on the value produced by a domain-specific evaluation function. In particular, a new solution is produced by using a random pair relocation method (see [39] for details), while the evaluation function is a lexicographic ordering function defined as:

$$e(\sigma) = \langle |\sigma|, -\sum_{r\in\sigma} |r|^2, \sum_{r\in\sigma} t(r)\rangle \tag{9}$$

where the first term is the number of routes, the second term tends to favor solutions with many customers and the last term takes into account the travel cost of the routing plan.

The second stage of the algorithm proposed in [39] minimizes the total travel cost by using a large neighborhood search (LNS) method. It consists of exploring the neighborhood of a given solution to find a better one, i.e. one that lowers the value of the objective function 8. We refer the reader to [39] for additional details on the above mentioned algorithms.

4 Experiments and Discussion

In this section, we show how the VRP optimization algorithm described above works presenting two examples. The common assumption for both examples is the use of a single depot (the same as deposit), a likely scenario where the teams of operators and their vehicles have a single starting location. We begin with a toy example in order both to show how the algorithm works and how we set up simulations; we then show a second subset of real scenarios with plants located in Sicily (south Italy) where some benefits of the proposed optimization strategy are discussed by taking into account both travel costs and worker teams.

4.1 Basic Setup

In the toy example, each pickup point corresponds to a single delivery, whilst in the second example multiple delivery points can share the same pickup site. Furthermore, we assumed to load the same quantity of spares material for each pickup and to choose the means of transport (van) so that it can contain the material necessary for all the delivery points. Finally, the cost (i.e. the distance or the time) between each pair of nodes is the same and it is set to 1.

This example is based on 10 nodes and 1 depot with a symmetric topology. In particular, this setup encompasses five plain routes connecting five places, called A, B, C, D, E, where each route contains only one pickup and one delivery point as shown in Fig. 1.

The problem is described by Eqs. (10), where $\mathcal{P}$, $\mathcal{D}$, $\mathcal{C}$ are the set of Pickups, Deliveries and Customers respectively, and by the set of routes described in Eqs. (11).

$$\mathcal{P} = \{A, B, C, D, E\}$$
$$\mathcal{D} = \{@A, @B, @C, @D, @E\}$$
$$\mathcal{C} = \mathcal{P} \cup \mathcal{D} = \{A, B, C, D, E, @A, @B, @C, @D, @E\} \tag{10}$$

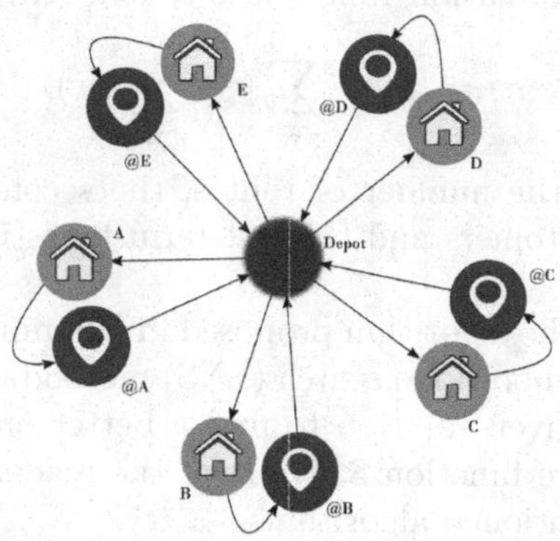

Fig. 1. The topology of the toy example

$$r_1 = \{depot, A, @A, depot\}$$
$$r_2 = \{depot, B, @B, depot\}$$
$$r_3 = \{depot, C, @C, depot\}$$
$$r_4 = \{depot, D, @D, depot\}$$
$$r_5 = \{depot, E, @E, depot\} \tag{11}$$

The algorithm attempts to optimize the solution according to the following two steps:

- SA: routes are reduced by using the Simulated Annealing algorithm.
- LNS: routes are optimized with the *Travel Cost Minimize* Function.

Note that the LNS step may not converge; when this happens, it is advisable to change some values of initial setup parameters and restart the algorithm from scratch. The values of the algorithm parameters are summarized in Table 1.

SA step looks for new routes with a better cost. In this example it finds the following four routes with a global cost of 14 (as shown in Fig. 2a).

$$r_0 = \{depot, A, @A, depot\}$$
$$r_1 = \{depot, B, @B, , depot\}$$
$$r_2 = \{depot, C, @C, depot\}$$
$$r_3 = \{depot, D, E, @E, @D, depot\} \tag{12}$$

Table 1. First example setting

Step	Item	Value
SA	Temperature value	28
	Temperature Limit value	15
	α	0.5
	Max Iterations	1
	β	1
LNS	Max Searches	5
	Max Iterations	1
	beta	2

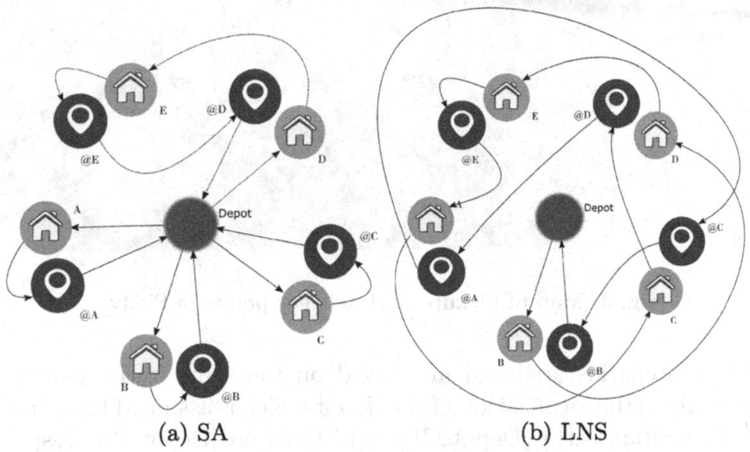

Fig. 2. Toy example: routes optimization.

Finally, the *LNS* optimization produces a single route 13, shown in Fig. 2b, whose cost is 10, that is much better than the initial value.

$$r_0 = \{depot, B, D, E, @E, A, C, @D, @A, @C, @B, depot\} \tag{13}$$

4.2 Real Case: Sicilian Wind Farms Scenarios

In this subsection we consider a real topology based on a set of wind farms located in Sicily. Three different scenarios are analyzed in this topology, showing how the adopted algorithm leads to cost savings together with the capability to provide fast recalculation of the routes to react to request changes.

As in the previous example, we have done some simplifications (e.g. we assume the same service time for each delivery point) while a more complex cost assessment model is used to take into account both the distances among the sites, the territory orography and the transportation network connecting the various sites.

The topology chosen uses a single depot (starting point for working teams) as in the toy example. Nine delivery points (wind turbines) and three pickups (spare parts storage) located in Sicily, Table 2 and Fig. 3 show the details of the topology.

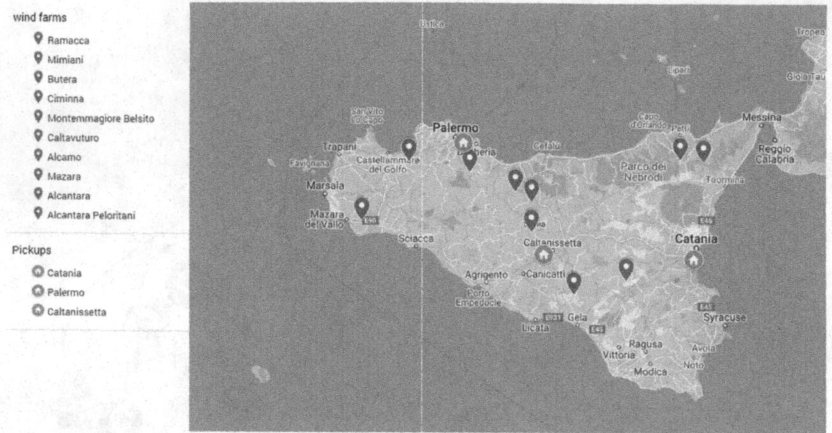

Fig. 3. Map of pickups and delivery points in Sicily

The three scenarios analyzed are based on the same initial configuration. They differ only in the localization of the depot which is assumed to be in Catania (Depot_CT), Caltanissetta (Depot_CL), and Palermo (Depot_PA) respectively.

Table 2. Pickup and delivery points

Pickup	Delivery
Catania	Alcantara Peloritani
	Alcantara
Palermo	Caltavuturu
	Ciminna
	Alcamo
	Mazara
Caltanissetta	Butera
	Mimiani
	Montemaggiore Belsito

Table 3 summarizes the initial configurations, i.e. both the parameters related to the service model and the simulation initial values. The former includes the loading capability and times need to load and to carry out the maintenance, which are supposed to be the same for all the scenarios. The latter includes the maximum number of iterations and the deepness of the *LNS* algorithm.

The initial setup (initial routes) is based on a simple on-request approach where each request is satisfied as soon as it is received; each team pickups the spare, travels to the destination, carries out both regular and extra maintenance at the client base, and returns to the starting point.

Table 3. Initial configuration

Parameter	Quantity
Pickup delivery time	15 min
Delivery service time	2 h and 30 min
Maximum loads	3 units
Downloaded units at delivery point	1 unit
SA max iterations	10 000
LNS max searches	2
LNS max iterations	5 000

The cost of the travel between two locations depends on several factors that have a different impact on the optimization strategy. The company policy may indeed aim at minimizing the intervention time, the cost of fuel, the personnel expenses, or any combination of them. A large van (Q = 3.5 tons) equipped as an electro-mechanical workshop and capable of holding up to 3 load units is used for maintenance crew and materials transportation. The estimated consumption is about 7.1 l/100 km. We minimized the travel time and then others costs were introduced to evaluate the effectiveness of the result. Indeed, the distances and travel time were calculated using Google map service, thus leading to a real travel time, that is generally lower between two distant locations connected by a highway, and higher for closer sites poorly connected.

All the simulated scenarios encompass the routes needed to cover both ordinary and opportunistic maintenance on the nine wind farms and three pickups. The aim is to reduce the number of worker teams and the global costs. The estimated average intervention time for a team of three electro-mechanical workers is 2.5 h for each visit.

The first scenario considered (Depot_CT) places the depot in Catania and the initial planning consists of four routes:

1. Depot_CT, Catania, Alcantara Peloritani, Catania, Alcantara, Depot_CT,
2. Depot_CT, Palermo, Caltavuturu, Palermo, Ciminna, Depot_CT
3. Depot_CT, Palermo, Alcamo, Palermo, Mazara, Depot_CT
4. Depot_CT, Caltanissetta, Butera, Caltanissetta, Mimiani, Caltanissetta, Montemaggiore Belsito, Depot_CT.

However, a complete study of the scenarios must include some other indicators to better estimate the cost of each route. In particular, to evaluate the monetary costs of ordinary maintenance, we have to take into account the overall cost of the process (following values come from standard 2019 italian labor market agreements):

1. The cost of the worker team, that is 68,90 €/hour, assuming each team is composed by 2 workers and 1 skilled technician;
2. the cost of running a van is supposed to be 7,1 €/100 km (avg. consumption);
3. the cost of loading and unloading of the van is 20,30 €/h.

Table 4. Cost of trivial case of scenario CT

Id	Depot	Pickups	Time (hours)			Cxost (euros)		
			Travel	Interv.	Total	Workers	Vans	Total
Depot_CT	Catania	2	6:16	5:30	11:46	810.72	33.23	843.95
		2	8:08	5:30	13:38	939.34	60.92	1 000.25
		2	9:48	5:30	15:18	1 054.17	73.47	1 127.64
		3	7:59	8:15	16:14	1 118.48	46.06	1 164.53
								4 136.38

Table 4 reports the costs of the original routes (trivial case) of the scenario Depot_CT. In this scenario, the simulated annealing phase converges into two routes, while the Large Neighborhood Search stage is not able to find better results, thus the final result is:

1. Depot_CT, Catania, Alcantara Peloritani, Catania, Alcantara, Depot_CT
2. Depot_CT, Palermo, Caltanissetta, Caltanissetta, Butera, Caltanissetta, Montemaggiore Belsito, Mimiani, Alcamo, Palermo, Palermo, Palermo, Caltavuturu, Ciminna, Mazara, Depot_CT

The results of simulation related to depot in Caltanissetta (scenario Depot_CL), whose costs in trivial case are reported in Table 5, are:

1. Depot_CL, Catania, Caltanissetta, Caltanissetta, Butera Caltanissetta, Alcantara Peloritani, Catania, Mimiani, Montemaggiore Belsito, Alcantara, Depot_CL
2. Depot_CL, Palermo, Palermo, Palermo, Ciminna, Alcamo, Palermo, Caltavuturu, Mazara, Depot_CL.

These results are similar to those found in the scenario Depot_CT; also in this case we found two routes during the simulated annealing stage and *LNS* is not able to further improve these results. Indeed, the resulting routes involves significantly less distance and travel time with a minor number of working teams.

Considering the depot located in Palermo (scenario Depot_PA), with Table 6 reporting initial costs, the simulated annealing stage was not able to reduce the initial routes, but the Large Neighborhood Search algorithm achieves the best results both in distance and duration. The routes found by LNS are as follows:

Table 5. Cost of trivial case of scenario CL

Id	Depot	Pickups	Time (hours)			Cost (euros)		
			Travel	Interv.	Total	Workers	Vans	Total
Depot_CL	Caltanissetta	2	8:49	5:30	14:19	986.42	52.06	1 038.48
		2	6:25	5:30	11:55	821.06	39.78	860.84
		2	7:20	5:30	10:26	884.22	47.81	932.03
		3	5:58	8:15	14:13	979.53	28.71	1 008.23
								3 839.58

Table 6. Cost of trivial case of scenario PA

Id	Depot	Pickups	Time (hours)			Cost (euros)		
			Travel	Interv.	Total	Workers	Vans	Total
Depot_PA	Palermo	2	9:15	5:30	14:45	1 016.28	60.92	1 077.19
		2	3:16	5:30	8:46	604.02	21.14	625.16
		2	4:56	5:30	10:26	718.86	33.32	752.18
		3	7:21	8:15	15:36	1 074.84	39.97	1 114.81
								3 569.34

1. Depot_PA, Catania, Catania, Alcantara, Alcantara Peloritani, Depot_PA
2. Depot_PA, Palermo, Caltanissetta, Montemaggiore Belsito, Palermo, Caltanissetta, Alcamo, Palermo, Mimiani, Caltavuturu, Palermo, Caltanissetta, Butera, Ciminna, Mazara, Depot_PA.

In Table 7 we summarize the results obtained with the SA and LNS in the three considered scenarios and Table 8 summarizes the costs of each scenario.

Table 7. Results of the 3 Scenarios

Scenario	Depot	Route (#)	Distance (km)	Travel time (hours)
Depot_CT	Catania	initial (4)	2 315	32:11
		after SA (2)	1 824	26:06
Depot_CL	Caltanissetta	initial (4)	1 824	28:49
		after SA (2)	1 459	27:21
Depot_PA	Palermo	initial (4)	1 488	24:00
		after SA (3)	2 436	32:54
		after LNS (2)	2038	28:40

The costs depend mainly on the number of teams. While in both trivial strategies each team starts from the pickup and came back after the intervention, and all routes of initial planning exhibit comparable length, the optimization provides better results with only two teams in any case.

Table 8. Evaluation of monetary cost

Id	Depot	Pickups	Time (hours)			Cost (euros)		
			Travel	Interv.	Total	Workers	Vans	Total
Depot_CT	Catania	2	6:14	5:30	11:44	808.43	33.23	841.65
		7	19:52	19:15	39:07	2 695.14	135.13	2 830.27
								3 671.92
Depot_CL	Caltanissetta	5	17:15	13:45	31:00	2 135.90	105.87	2 241.77
		4	10:06	11:00	21:06	1 453.79	68.03	1 521.82
								3 763.58
Depot_PA	Palermo	2	7:23	5:30	12:53	887.66	49.10	936.77
		7	21:17	19:15	40:32	2 792.75	139.00	2 931.75
								3 868.52

The results also show that costs are strongly dependent on the manpower rather than on the location of the depot.

Moreover, the proposed approach permits to rapidly change the work plan whenever needed and without the support of a manager, achieving an additional cost saving. Each run presented in this section was indeed executed on a Desktop PC with a standard configuration (16 GB memory Intel i5 processor) in about 20 min in the worst case.

Table 9. Maintenance management KPI comparison

KPI	Definition	depot_CT		Variation
		trivial	best	
A&S7	Total internal personnel cost spent in maintenance/Total Maintenance Cost	79%	67%	−12%
A&S11	Cost of indirect maintenance personnel/Total maintenance cost	7%	6%	−1%
PHA8	Preventive maintenance time causing downtime/Total downtime related to maint	74%	83%	12%
E12	Total Operating time/(Total Operating time + Downtime due to maintenance)	33%	50%	17%
O&S22	Preventive maintenance man-hours/Total maintenance man-hours	30%	40%	10%

Comparing the trivial configurations with the analogous scenario using the algorithm, there is a reduction of both crew working on field and of indirect personnel, an increasing in work-orders technically completed in the correct time-window, and a general increase of total operating time. Also through UNI EN 15341:2918 KPIs, as shown in Table 9 only for the scenario having Catania as

Depot (Depot_CT), it is possible to appreciate the significant savings obtained applying the proposed algorithm on costs of internal (A&S7) and indirect maintenance personnel (A&S11), a general increase of time and man-hours for preventive maintenance (PHA8 and O&S22) which have a positive effect on total operating time (E12) due to a reduced downtime for corrective maintenance and a more efficient maintenance scheduling.

5 Conclusions and Future Works

In this work we discussed a PDPTW algorithm implementation to optimize maintenance intervention routing in wind farms, considering overall costs of the intervention based on human resources and fuel consumption. After a simple toy example, a real scenario of wind farms in Sicily was considered with different depot placement and corresponding costs assessment. Results show the effectiveness of the proposed approach and more efficient use of maintenance man-hours.

Indeed, comparing the trivial configuration with the analogous scenario having Catania as Depot (Depot_CT) there is a reduction of both crew working on the field and of indirect personnel, an increase in work-orders technically completed within the correct time-window, and a general increase of total operating time. These results can be appreciated also observing five KPI from UNI EN 15341:2019 which refers to administration and procurement subfunction, maintenance for asset management, maintenance engineering subfunction, organization and support subfunction. Their values show significant improvement demonstrating that the algorithm can successfully support maintenance management decisions.

In addition, the PDPTW implementation discussed here allows new configuration of the wind farms maintenance management more proactively via preventive, condition-based and opportunistic maintenance rather than operating with corrective maintenance only.

In summary, the contribution of this paper mainly relies in highligthing the effectiveness and efficiency of VRP algorithm for wind farms maintenance optimization. The proposed approach also exhibits a couple of limitations, as

- the period considered for experiments is quite short;
- real scenario considered is limited to a single region (Sicily);
- the cost assessment subfunction does not include KPIs.

To overcome these issues as well,
future directions include the following:

- the proposed approach should be adopted for a longer period to consolidate short-time (although encouraging) results presented above;
- more complex scenario, for instance considering all wind farms in Italy, should be investigated to strengthen cost assessment;
- the PDPTW algorithm should be integrated within standard operational processes, to improve short as well as long term activity plan;

- the cost assessment subfunction should be improved by taking into account KPIs;
- just-in-time optimization can be considered to tackle with unplanned requests that might arise during interventions (e.g. harsh weather).

Acknowledgment. This work has been partially supported by the project of University of Catania PIACERI, *PIAno di inCEntivi per la Ricerca di Ateneo.*

References

1. Castellani, F., Astolfi, D., Sdringola, P., Proietti, S., Terzi, L.: Analyzing wind turbine directional behavior: SCADA data mining techniques for efficiency and power assessment. Appl. Energy **185**, 1076–1086 (2017). https://doi.org/10.1016/j.apenergy.2015.12.049
2. Merkt, O.: Predictive models for maintenance optimization: an analytical literature survey of industrial maintenance strategies. In: Ziemba, E. (ed.) AITM/ISM -2019. LNBIP, vol. 380, pp. 135–154. Springer, Cham (2020). https://doi.org/10.1007/978-3-030-43353-6_8
3. Ferreira, R.S., Feinstein, C.D., Barroso, L.A.: Operation and maintenance contracts for wind turbines. In: Sanz-Bobi, M.A. (ed.) Use, Operation and Maintenance of Renewable Energy Systems. GET, pp. 145–181. Springer, Cham (2014). https://doi.org/10.1007/978-3-319-03224-5_5
4. Zhang, X., Zeng, J.: A general modeling method for opportunistic maintenance modeling of multi-unit systems. Reliab. Eng. Syst. Saf. **140**, 176–190 (2015). https://doi.org/10.1016/j.ress.2015.03.030
5. Dai, J., Yang, W., Cao, J., Liu, D., Long, X.: Ageing assessment of a wind turbine over time by interpreting wind farm SCADA data. Renew. Energy **116**, 199–208 (2018). https://doi.org/10.1016/j.renene.2017.03.097
6. Jin, X., Xu, Z., Qiao, W.: Condition monitoring of wind turbine generators using SCADA data analysis. IEEE Trans. Sustain. Energy **99**, 1 (2020). https://doi.org/10.1109/TSTE.2020.2989220
7. Chen, L., Xu, G., Zhang, Q., Zhang, X.: Learning deep representation of imbalanced SCADA data for fault detection of wind turbines. Measurement **139**, 370–379 (2019). https://doi.org/10.1016/j.measurement.2019.03.029
8. Dao, P.B., Staszewski, W.J., Barszcz, T., Uhl, T.: Condition monitoring and fault detection in wind turbines based on co-integration analysis of SCADA data. Renew. Energy **116**, 107–122 (2018). https://doi.org/10.1016/j.renene.2017.06.089
9. Bangalore, P., Patriksson, M.: Analysis of SCADA data for early fault detection, with application to the maintenance management of wind turbines. Renew. Energy **115**, 521–532 (2018). https://doi.org/10.1016/j.renene.2017.08.073
10. Carchiolo, V., Longheu, A., Di Martino, V., Consoli, N.: Power plants failure reports analysis for predictive maintenance. In: Proceedings of the 15th International Conference on Web Information Systems and Technologies (WEBIST), vol. 1, pp. 404–410. INSTICC, SciTePress (2019). https://doi.org/10.5220/0008388204040410
11. Leyh, C., Martin, S., Schäffer, T.: Analyzing industry 4.0 models with focus on lean production aspects. In: Ziemba, E. (ed.) AITM/ISM-2017. LNBIP, vol. 311, pp. 114–130. Springer, Cham (2018). https://doi.org/10.1007/978-3-319-77721-4_7

12. Carchiolo, V., Catalano, G., Malgeri, M., Pellegrino, C., Platania, G., Trapani, N.: Modelling and optimization of wind farms' processes Using BPM. In: Ziemba, E. (ed.) AITM/ISM -2019. LNBIP, vol. 380, pp. 95–115. Springer, Cham (2020). https://doi.org/10.1007/978-3-030-43353-6_6
13. Carchiolo, V., et al.: Pick-up & delivery in maintenance management of renewable energy power plants. In: 15th Conference on Computer Science and Information Systems (FedCSIS), pp. 579–585 (2020). https://doi.org/10.15439/2020F114
14. Perez-Canto, S., Rubio-Romero, J.C.: A model for the preventive maintenance scheduling of power plants including wind farms. Reliab. Eng. Syst. Saf. **119**, 67–75 (2013). https://doi.org/10.1016/j.ress.2013.04.005
15. Yang, L., Li, G., Zhang, Z., Ma, X., Zhao, Y.: Operations maintenance optimization of wind turbines integrating wind and aging information. IEEE Trans. Sustain. Energy **12**(1), 211–221 (2021). https://doi.org/10.1109/TSTE.2020.2986586
16. Lopes, R.S., Cavalcante, C.A., Alencar, M.H.: Delay-time inspection model with dimensioning maintenance teams: a study of a company leasing construction equipment. Comput. Indus. Eng. **88**, 341–349 (2015). https://doi.org/10.1016/j.cie.2015.07.009
17. Si, G., Xia, T., Zhu, Y., Du, S., Xi, L.: Triple-level opportunistic maintenance policy for leasehold service network of multi-location production lines. Reliab. Eng. Syst. Saf. **190**, 106519 (2019). https://doi.org/10.1016/j.ress.2019.106519
18. Raza, A., Ulansky, V.: Optimal preventive maintenance of wind turbine components with imperfect continuous condition monitoring. Energies **12**(19), 3801 (2019). https://doi.org/10.3390/en12193801
19. Kang, J., Guedes Soares, C.: An opportunistic maintenance policy for offshore wind farms. Ocean Eng. **216**, 108075 (2020). https://doi.org/10.1016/j.oceaneng.2020.108075
20. Li, M., Wang, M., Kang, J., Sun, L., Jin, P.: An opportunistic maintenance strategy for offshore wind turbine system considering optimal maintenance intervals of subsystems. Ocean Eng. **216**, 108067 (2020). https://doi.org/10.1016/j.oceaneng.2020.108067
21. Shafiee, M., SÃžrensen, J.D.: Maintenance optimization and inspection planning of wind energy assets models methods and strategies. Reliab. Eng. Syst. Saf. **192**, 105993 (2019). https://doi.org/10.1016/j.oceaneng.2020.108075
22. Dalgic, Y., Lazakis, I., Dinwoodie, I., McMillan, D., Revie, M.: Advanced logistics planning for offshore wind farm operation and maintenance activities. Ocean Eng. **101**, 211–226 (2015). https://doi.org/10.1016/j.oceaneng.2015.04.040
23. Tan, Q., Wei, T., Peng, W., Yu, Z., Wu, C.: Comprehensive evaluation model of wind farm site selection based on ideal matter element and grey clustering. J. Clean. Prod. **272**, 122658 (2020). https://doi.org/10.1016/j.jclepro.2020.122658
24. Yan, B., Ma, Y., Zhou, Y.: Research on spare parts inventory optimization in wind power industry. In: 2020 Global Reliability and Prognostics and Health Management (PHM-Shanghai), pp. 1–5 (2020). https://doi.org/10.1109/PHM-Shanghai49105.2020.9280922
25. Liu, R., Dan, B., Zhou, M., Zhang, Y.: Coordinating contracts for a wind-power equipment supply chain with joint efforts on quality improvement and maintenance services. J. Clean. Prod. **243**, 118616 (2020). https://doi.org/10.1016/j.jclepro.2019.118616
26. Santos, M., González, M.: Factors that influence the performance of wind farms. Renew. Energy **135**, 643–651 (2019). https://doi.org/10.1016/j.renene.2018.12.033

27. Nguyen, T.A.T., Chou, S.Y.: Improved maintenance optimization of offshore wind systems considering effects of government subsidies, lost production and discounted cost model. Energy **187**, 115909 (2019). https://doi.org/10.1016/j.energy.2019.115909
28. Zhu, W., Castanier, B., Bettayeb, B.: A dynamic programming-based maintenance model of offshore wind turbine considering logistic delay and weather condition. Reliab. Eng. Syst. Saf. **190**, 106512 (2019). https://doi.org/10.1016/j.ress.2019.106512
29. UNI, EN: 15341:2019 Maintenance - Maintenance Key Performance Indicators. http://store.uni.com/catalogo/uni-en-15341-2019
30. Zhong, S., Pantelous, A.A., Goh, M., Zhou, J.: A reliability-and-cost-based fuzzy approach to optimize preventive maintenance scheduling for offshore wind farms. Mech. Syst. Signal Process. **124**, 643–663 (2019). https://doi.org/10.1016/j.ymssp.2019.02.012
31. Yurusen, N.Y., Rowley, P.N., Watson, S.J., Melero, J.J.: Automated wind turbine maintenance scheduling. Reliab. Eng. Syst. Saf. **200**, 106965 (2020). https://doi.org/10.1016/j.ress.2020.106965
32. Fan, D., Ren, Y., Feng, Q., Zhu, B., Liu, Y., Wang, Z.: A hybrid heuristic optimization of maintenance routing and scheduling for offshore wind farms. J. Loss Prev. Process Indus. **62**, 103949 (2019). https://doi.org/10.1016/j.jlp.2019.103949
33. Irawan, C.A., Eskandarpour, M., Ouelhadj, D., Jones, D.: Simulation-based optimisation for stochastic maintenance routing in an offshore wind farm. Eur. J. Oper. Res. **289**(3), 912–926 (2021). https://doi.org/10.1016/j.ejor.2019.08.032
34. Gutierrez-Alcoba, A., Hendrix, E., Ortega, G., Halvorsen-Weare, E., Haugland, D.: On offshore wind farm maintenance scheduling for decision support on vessel fleet composition. Eur. J. Oper. Res. **279**(1), 124–131 (2019). https://doi.org/10.1016/j.ejor.2019.04.020
35. Kovács, A., Erdös, G., Viharos, Z.J., Monostori, L.: A system for the detailed scheduling of wind farm maintenance. CIRP Ann. **60**(1), 497–501 (2011). https://doi.org/10.1016/j.cirp.2011.03.049
36. Froger, A., Gendreau, M., Mendoza, J.E., Pinson, E., Rousseau, L.M.: A branch-and-check approach for a wind turbine maintenance scheduling problem. Comput. Oper. Res. **88**, 117–136 (2017). https://doi.org/10.1016/j.cor.2017.07.001
37. Jbili, S., Chelbi, A., Radhoui, M., Kessentini, M.: Integrated strategy of vehicle routing and maintenance. Reliab. Eng. Syst. Saf. **170**, 202–214 (2018). https://doi.org/10.1016/j.ress.2017.09.030
38. Sasmi Hidayatul, Y.T., Djunaidy, A., Muklason, A.: Solving multi-objective vehicle routing problem using hyper-heuristic method by considering balance of route distances. In: 2019 International Conference on Information and Communications Technology (ICOIACT), pp. 937–942 (2019). https://doi.org/10.1109/ICOIACT46704.2019.8938484
39. Bent, R., Hentenryck, P.V.: A two-stage hybrid algorithm for pickup and delivery vehicle routing problems with time windows. Comput. Oper. Res. **33**(4), 875–893 (2006). https://doi.org/10.1016/j.cor.2004.08.001. part Special Issue: Optimization Days 2003

An Improved Map Matching Algorithm Based on Dynamic Programming Approach

Alexander Yumaganov, Anton Agafonov(✉), and Vladislav Myasnikov

Samara National Research University,
34, Moskovskoye shosse, Samara 443086, Russia
yumagan@gmail.com, ant.agafonov@gmail.com, vmyas@geosamara.ru

Abstract. GPS sensors embedded in almost all mobile devices and vehicles generate a large amount of data that can be used in both practical applications and transportation research. Despite the high accuracy of location measurements in 3–5 m on average, this data can not be used for practical use without preprocessing. The preprocessing step that is needed to identify the correct path as a sequence of road segments by a series of location measurements and road network data is called map matching. In this paper, we consider the offline map matching problem in which the whole trajectory is processed after it has been collected. We propose a map matching algorithm based on a dynamic programming approach. To enhance the quality of the map matching algorithm, we propose a modification of the algorithm aimed at enhancing the accuracy of the map matching procedure. The modification of the algorithms consists in dividing the GPS trajectory into sections and step-by-step running the base map matching algorithm for each section. The experimental studies were conducted on the dataset collected in Samara, Russia, and the publicly available large-scale dataset for testing, benchmarking, and offline learning of map matching algorithms. Experiments showed that the proposed algorithm outperforms other comparable algorithms in terms of accuracy.

Keywords: Map matching · GPS data · Dynamic programming

1 Introduction

Widespread deployment of the Global Positioning System (GPS) provides a large amount of data describing movement trajectories of pedestrians, bicycles, vehicles, etc. The trajectories are observed as a sequence of GPS records. Each record usually contains ID, latitude, and longitude of the GPS sensor and timestamp of the record. GPS sensors usually provide location data with high accuracy up to 5 m on average, but in some cases, the measurement errors can be much

The work was partially supported by RFBR research projects no. 18-29-03135-mk.

E. Ziemba and W. Chmielarz (Eds.): ISM 2020/FedCSIS-IST 2020, LNBIP 413, pp. 87–102, 2021.
https://doi.org/10.1007/978-3-030-71846-6_5

higher, especially in the urban environment. In any case, to use the GPS data in many practical applications and transportation research we first need to perform the preprocessing step that is called the map matching process. Map matching algorithms are applied to identify the correct path as a sequence of road segments by a series of location measurements (GPS records) and road network data. Processed trajectories are an important data source for intelligent transportation systems that can be used in such applications as traffic estimation and prediction [1,2], traffic modeling [3], developing navigation services [4], solving logistic problems [5,6], user preferences elicitation and training of transportation recommendation systems [7,8], and so on.

As mentioned earlier, the GPS-trajectories have measurement errors because of multiple factors: atmospheric phenomena, interference from ground-based radio sources, high-rise urban development, vegetation, imperfect hardware and the embedded processing algorithms, and others. We considered a large number of GPS trajectories collected by several mobile devices and identify several typical errors:

1. Large geolocation error. The GPS sensor gives several records with the measurement error significantly higher than the usual 3–5 m.
2. Large time gaps. The GPS sensor does not provide any data for a long time. There are gaps in the trajectory in several minutes or more between high-quality recorded fragments.
3. Continuous deviations from the ground truth path. The GPS sensor for a long time provides coordinates with a low error, the trajectory looks smooth, however, the deviation of the trajectory from the true path is several times higher than the average measurement error.
4. Loops when stopped after fast movement. The GPS sensor provides coordinates with a low error, however, the trajectory forms smooth walks in various directions and loops. It is noticed that this happens on any device, but under the same conditions: during a sudden stop after a fast movement, for example, at a traffic light.

Examples of the described errors are plotted in Fig. 1. The ideal path is shown by the black line and the GPS trajectory is plotted by the red-dotted line.

Given the above measurement errors, the map matching process can be quite challenging. As a result, a number of map matching algorithms have been developed to solve this problem.

The map matching algorithms can be categorized by different criteria. In this paper, we consider the online/offline classification. Online map matching methods [9–11] process positions when the trajectory is still collecting. Offline map matching methods [12,13] compute the path after the whole trajectory has been collected. In this paper, we consider the offline map matching problem. A detailed literature review is presented in the next section.

The rest of the paper is organized as follows. Related work is reviewed in Sect. 2. In Sect. 3, we give the main notation and problem statement, describe the base map matching algorithm with the dynamic algorithm of the shortest path assessment, and present an algorithm modification aimed at improving the

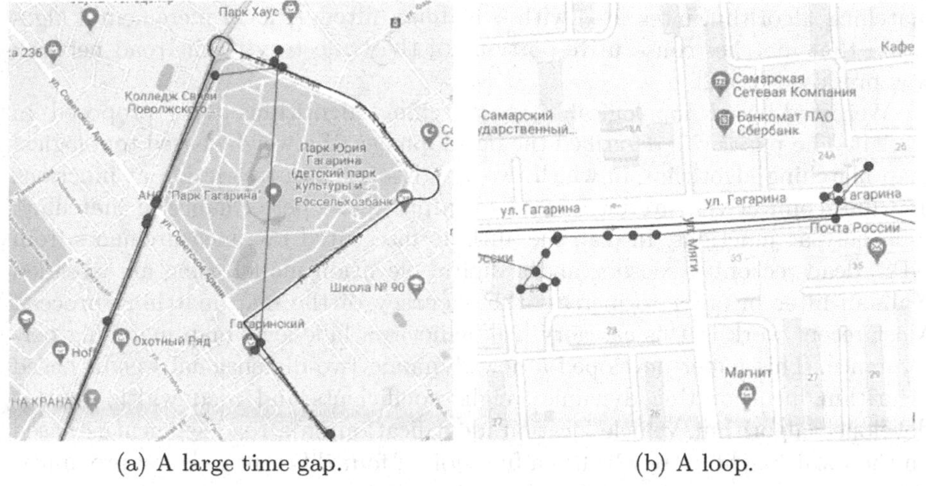

(a) A large time gap. (b) A loop.

Fig. 1. GPS measurement errors. (Color figure online)

accuracy of the algorithm. Section 4 describes the experimental setup and results of experimental studies. Finally, we give a conclusion and possible directions for further research.

2 Literature Review

Classification and comparative study of map matching algorithms were presented in [14]. In [15], the authors reviewed existing map matching algorithms with the aim of highlighting their qualities, unresolved issues, and provide directions for future studies. The algorithms were compared with respect to positioning sensors, map qualities, assumptions, and accuracy. A new categorization of the map matching solutions according to their map matching models and working scenarios was proposed in [16]. In addition, the authors experimentally compare three representative methods from different categories to reveal how the matching model affects the performance.

Simple geometrical-based map matching algorithms include point-to-point, point-to-curve and curve-to-curve methods [10,17,18]. Geometrical algorithms project each trajectory point or each trajectory segment to the geometric-closest edge. However, these algorithms are sensitive to trajectory measurement errors and often fail in complex urban road networks [19]. Topological map matching algorithms compare the geometrical and topological characteristics of the road network with the characteristics of the vehicle's trajectory [10,20]. Basic map matching methods were investigated in [10]. The authors considered a classical point-to-curve algorithm, modified versions where vehicle heading and road-network topology are taken into consideration, and the curve-to-curve matching algorithm. In [21], the authors developed a topological point-to-curve map matching algorithm integrated with a Kalman filter. A local incremental algorithm that matches consecutive portions of the trajectory to the road network was proposed in [22].

Weighted-based topological map matching algorithms were proposed in [23,24]. The paper [25] described the development of a weight-based topological map matching algorithm, in which two weights for turn-restriction at junctions and link connectivity are introduced to improve the performance of matching, especially at junctions. In [24], the authors integrated raw measurements from GPS, dead-reckoning sensors, and a digital elevation model using an extended Kalman filter in order to increase the accuracy of the map matching process. Most recent work in this category [26] achieves a lane-level map matching performance. This paper developed a new dynamic two-dimensional weight-based algorithm incorporating dynamic weight coefficients and road width. In this developed algorithm, vehicle location identification on a road segment is based on the total weight score which is a function of four different weights: proximity, kinematic, turn-intent prediction, and connectivity.

In later works, advanced map matching algorithms was proposed. In [27], the authors proposed a map matching algorithm for large-scale low-frequency floating car data. They used a multicriteria dynamic programming technique to minimize the number of candidate routes for each GPS point. The paper [28] proposed a fuzzy logic based algorithm to map match data from a high sensitivity GPS and a gyroscope. In [29], the authors developed a map matching algorithm based on fuzzy logic theory. The inputs to the proposed algorithm are from GPS augmented with data from deduced reckoning sensors to provide continuous navigation. In [30,31], the authors discussed the possibility of applying Hidden Markov models (HMM) in map matching algorithms. In the proposed methods, the authors used HMM to find the most likely road route taking into account the measurement noise and the layout of the road network. In [32], the authors proposed a feature-based map matching algorithm that estimates the cost of a candidate path based on both GPS observations and a behavioral model. A map matching algorithm based on Dijkstra's shortest path method that is applicable for large scale datasets was proposed in [12]. The authors focused on reducing the computational complexity of the algorithm. In [33], the authors also concentrated on designing efficient and scalable map matching algorithms. They presented an algorithm integrating the hidden Markov model with precomputations of all shortest path pairs within a certain length in the road network. This allowed replacing the routing queries with hash table search. A map matching based framework to reconstruct vehicular trajectories from GPS datasets was presented in [34]. The authors proposed a framework that consists of four components: preprocessing, candidate set, information modeling, and solution construction. To reconstruct the trajectory from the sequence of candidates, two approaches were used: HMM-based construction and Ant Colony-based construction.

Despite the large number of papers devoted to the map matching problem, the proposed solutions do not allow achieving high accuracy or, in some cases, can not find the correct path at all. In this paper, we focus on developing the map matching algorithm that allows us to identify the correct path with high accuracy. The proposed algorithm consists of two steps: calculating the shortest paths and estimating the paths using a dynamic programming approach.

3 Research Methodology

3.1 Problem Statement

A road network is represented as a directed graph $G = (V, W)$, where V is the set of nodes that represent road intersections, W is the set of edges denote road segments. Each node $v \in V$ has the coordinates $\overline{x}_v = (x_v, y_v)$. Each edge $w_{ij} \in W, i, j \in V$ is described by the tuple:

$$w_{ij} = (l^w, v^w_{max}, X^w)\,, \tag{1}$$

where l^w is the length of the road segment w, v^w_{max} is the maximum allowed speed, X^w is the geometry of the road segment w presented as a set of points.

Define a GPS trajectory as the set of GPS records obtained during the observation:

$$\{\overline{x}_i, t_i\}_{i=\overline{0,I-1}}\,, \tag{2}$$

where I is the number of GPS records, $\overline{x}_i = (x_i, y_i)$ is the coordinates of the tracked objects (latitude and longitude), t_i is the timestamp of i-th GPS record.

Define the ground truth path P as the sequence of the edges (road segments) that were traversed by the vehicle during the observation.

Given the introduced notation, the map matching problem can be formulated as follows:

Given a graph $G = (V, W)$ and a GPS trajectory $\{\overline{x}_i, t_i\}_{i=\overline{0,I-1}}$ find the ground truth path P traversed by a vehicle in a road network.

3.2 Proposed Approach

Base Map Matching Algorithm. The base map matching algorithm can be described as a sequence of the following steps [35]:

1. Determine the start and end nodes by the coordinates of the first and the last GPS records.
2. For all edges $w \in W$ located at a distance to the GPS records not exceeding $R = 100\,\text{m}$, the edge weight is set using the following equation:

$$\varphi(w) = \left(1 - \frac{1}{K}\sum_{k=0}^{K-1} \exp\left(-\alpha\|\overline{x}_k - \overline{x}^p_k(w)\|^2\right)\right) l^w \tag{3}$$

where K is the number of GPS records matched with the edge w, $\overline{x}_k$ are the coordinates of the matched GPS record, $\overline{x}^p_k$ is the coordinates of the GPS record projection on the edge w, α is the coefficient.
For not matched edges the weight is set as follows:

$$\varphi(w) = \beta l^w, \tag{4}$$

where $\beta = 10$ is the coefficient.

3. In the graph with the edge weights set as described above, the shortest path is searched from the start to end node corresponding to the first and last GPS records.
4. The found shortest path is estimated using a dynamic algorithm described below.
5. The algorithm for sequential removal of edges from the path is performed. **Input data:** path, assessment of the path, graph. For an edge from the list of path edges:
   (a) The edge is removed from the graph.
   (b) The shortest path search from the start node of the removed edge to the end node is performed. If there is no such path in the graph, go to step e).
   (c) The resulting path is estimated using a dynamic algorithm.
   (d) If the resulting assessment is greater than the assessment of the original path, then save the resulting path to the list of best paths;
   (e) Restore the deleted edge and move on to process the next edge. **Output data:** the path from the list of best paths, select the path with the maximum assessment value, or an empty path if the list of best paths is empty.
6. The algorithm for sequential removal of edges runs in a loop until an empty path is returned. The last non-empty path will be a solution of the map matching algorithm.

In order to reduce the impact of typical errors in the GPS records, the path assessment in the algorithm for sequential removal of edges is calculated by the following quality criterion:

$$J^* = \begin{cases} J_{p\_cur} - \gamma \frac{l_{p\_cur} - l_{p\_base}}{J_{p\_cur} - J_{p\_base}}, J_{p\_cur} - J_{p\_base} > 0; \\ J_{p\_cur}, \, otherwise, \end{cases} \tag{5}$$

where J_{p_cur} is the current path assessment, J_{p_cur} is the base path assessment, l_{p_cur} is the current path length, l_{p_base} is the base path length, γ is an empirically selected coefficient.

In the next subsection, we describe the algorithm for the reconstructed path estimation.

Dynamic Algorithm for Reconstructed Path Estimation. As a criterion for the reconstruction quality, we use the following:

$$J_p = \sum_{i=0}^{I-1} \exp\left(-\alpha \|\overline{x}_i - \overline{x}_i^p\|^2\right). \tag{6}$$

We need to match the points $\{\overline{x}_i, t_i\}_{i=\overline{0,I-1}}$ with the path p. Firstly, we discretize the path into N points with the discretization step $\Delta = 2\,\text{m}$: $p(n) = \overline{x}_n^p, p = \overline{0, N-1}$. Next, we calculate I arrays of proximity similarities between the point $\overline{x}_i$ and the path p as:

$$\varphi_i(n) = \exp\left(-\alpha \|\overline{x}_i - p(n)\|^2\right). \tag{7}$$

The optimization problem is to find the sequence

$$n(i)_{i=\overline{0,I-1}} : \sum_{i=0}^{I-1} \varphi\left(n\left(i\right)\right) \rightarrow \max . \tag{8}$$

The main recurrence relation (for the dynamic programming algorithm) has the following form:

$$\begin{aligned} \max_{n(i)} \sum_{i=0}^{I-1} \varphi_i\left(n\left(i\right)\right) = \max_{n(i_l)=\overline{n(i_{l-1}),N}} \Bigg[\varphi_i\left(n\left(i\right)\right) \\ + \max_{n(i)\leq n(i_l)} \sum_{i=0}^{i_l-1} \varphi_i\left(n\left(i\right)\right) + \max_{n(i)\geq n(i_l)} \sum_{i=i_l-1}^{I-1} \varphi_i\left(n\left(i\right)\right) \Bigg] . \end{aligned} \tag{9}$$

Introduce the additional notations. Let $\widetilde{\varphi}_i(n)$ be the maximum integral similarity:

$$\widetilde{\varphi_j}(n) = \max_{n(i):i\leq j} \sum_{i=0}^{j} \varphi_i\left(n\left(i\right)\right). \tag{10}$$

Let $\pi_i(n)$ be the list of point positions.

The dynamic programming algorithm to solve the recurrence relation can be described as follows (Algorithm 1).

The result path assessment and the list of point positions are stored in $\widetilde{\varphi_0}(0)$ and $\pi_0(0)$. Using this path assessment, the best path by the specified criteria will be selected in the map matching algorithm.

The proposed algorithm has some limitations. To overcome this, in the next subsection, we propose modifications aimed at improving the quality of the map matching result.

Modified Map Matching Algorithm. The base version of the proposed map matching algorithm demonstrates poor matching quality in the following cases:

1. The route contains segments that have been visited multiple times during the movement along it.
2. The route contains segments that are located at a short distance relative to each other.

Figure 2 shows examples of cases with poor matching quality. The ground truth path is shown by the black line and the GPS trajectory is plotted by the red-dotted line.

We present a modification of the proposed map matching algorithm that improves the quality of matching in the described problem cases. The modification of the algorithm consists in dividing the GPS trajectory into sections and step-by-step running the base map matching algorithm (described in Subsect. 3.2) for each section. The modified version of the map matching algorithm consists of the following steps:

Algorithm 1. Path estimation algorithm

```
for i = I − 1, 0 do
  for n = N − 1, 0 do
    if i == I − 1 then
      if n == N − 1 then
        φ̃_i(N − 1) = φ_i(N − 1);
        π_i(N − 1) = {N − 1};
      else
        if φ_i(n) > φ̃_i(n + 1) then
          φ̃_i(n) = φ_i(n);
          π_i(n) = {n};
        else
          φ̃_i(n) = φ̃_i(n + 1);
          π_i(n) = π_i(n + 1);
        end if
      end if
    else
      if n == N − 1 then
        φ̃_i(N − 1) = φ_i(N − 1) + φ̃_{i+1}(N − 1);
        π_i(N − 1) = π_{i+1}(N − 1);
        π_i(N − 1).add(N − 1);
      else
        if φ_i(n) + φ̃_{i+1}(n) > φ̃_i(n + 1) then
          φ̃_i(n) = φ_i(n) + φ̃_{i+1}(n);
          π_i(n) = copy(π_{i+1}(n));
          π_i(n).add(n);
        else
          φ̃_i(n) = φ̃_i(n + 1);
          π_i(n) = π_i(n + 1);
        end if
      end if
    end if
  end for
end for
```

1. The trajectory points are filtered. Filtering removes points from the GPS trajectory located at a distance of less than $R_f = 20\,\text{m}$ to each other. As a result, the filtered GPS trajectory contains J points ($J \leq I$). Filtering is necessary to exclude closely located points from consideration when choosing the best matching edges for split points (a split point is defined below).
2. An initial matching of trajectory points to the edges of the road network graph is performed. As a result, for each point of the trajectory, the nearest edge of the graph is matched.
3. The GPS trajectory is divided into sections by split points $\{x_s\} \subset \{x_i\}$, $i = \overline{0, J-1}$, $s \leq J - 1$. A trajectory point x_i is a split point in the following cases:
   (a) x_i is the start or the end point of the trajectory ($i = 0 \vee i = J - 1$).

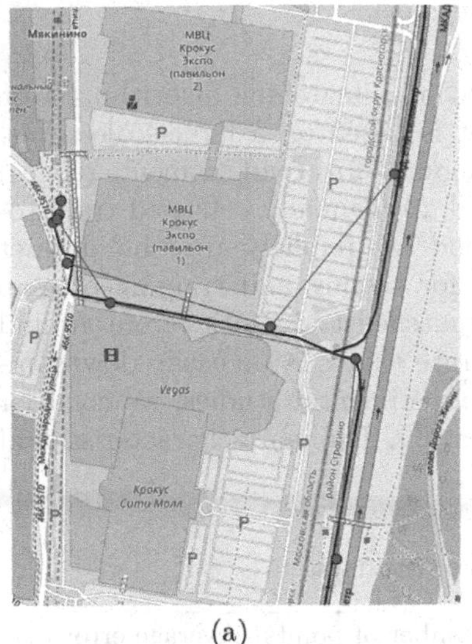

(a)

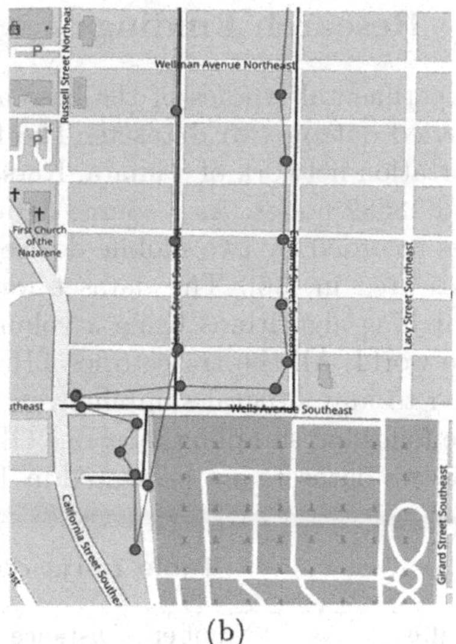

(b)

Fig. 2. Examples of the trajectories with map matching problems.

(b) The distance from the current split point (the closest preceding one to the i-th trajectory point) to the i-th trajectory point is greater than or equal to the distance from the current split point to the $i+1$ trajectory point.
(c) The distance from the current split point to the i-th trajectory point is less than the distance from the current split point to the $i+1$ trajectory point and the angle between the vectors formed by the coordinates of the current split point and i and $i+1$ trajectory points is more than 120°.

4. The split points are mapped on the road segments. To match the split points, we develop an algorithm based on the assumption that the initial mapping of most GPS records is performed correctly.
To match the split points, we develop an algorithm based on the assumption that the initial matching of most GPS records is performed with acceptable accuracy. Let us consider this algorithm in more detail. For each point from the set of split points $\{x_s\}$ we perform the following steps:
(a) Choose K pairs of trajectory points equidistant from the considered split point.
(b) Find the shortest path between each pair of the selected points.
(c) Match the split point to the nearest edge that is contained in the largest number of shortest paths.

Finally, the base version of the proposed map matching algorithm is sequentially applied to each of the trajectory sections.

4 Research Findings

Experimental studies of the base and modified map matching algorithms were carried out on two datasets. The first dataset is based on a large-scale transportation network of Samara, Russia, consisting of 47274 road segments (edges) and 18582 nodes. As a source dataset, we used 20 manually collected trajectories recorded by two mobile devices. The second is a publicly available dataset presented in [36]. The dataset was collected for testing and training of map matching algorithms using a volume of map-matched trajectories from around the world. All the trajectories 1 Hz sampling rate that can be easily downsampled to any integral sampling rate. In these experimental studies, we consider the following sampling rate: one GPS point per 30 s. The main characteristics of the two datasets are presented in Table 1.

Table 1. The characteristics of the datasets

Dataset	Number of trajectories	Distance per trajectory, km			Number of points per trajectory			Average error per trajectory, m		
		Min	Max	Avg	Min	Max	Avg	Min	Max	Avg
Samara dataset	20	2.65	11.01	6.61	24	968	438	1.45	215.1	10.69
Public dataset	100	5.02	94.35	26.8	10	644	83	3.41	5.69	9.29

We compare the proposed modified dynamic-based algorithm (DBA modified) with the base dynamic-based algorithm (DBA base), the FMM algorithm [33], and the HMM-based algorithm [30] implemented as a part of the GraphHopper library [37].

To evaluate the map matching accuracy, we used two metrics: Route Mismatch Fraction (RMF) introduced in [30] and the accuracy metric (A) that was used in [33].

The RMF is computed as:

$$RMF = \frac{1}{M}\sum_{m=0}^{M-1}\frac{l_{+}^{m}+l_{-}^{m}}{l_{gt}^{m}}, \tag{11}$$

where M is the number of trajectories, l_{gt}^{m} is the m-th ground truth path length, l_{+} is the length of the road segments in the m-th matched path that are not in the m-th ground truth path (erroneously added), l_{-} is the length of the road segments in the m-th ground truth path that are not presented in the m-th matched path (erroneously subtracted).

The accuracy metric is the average of the overlapping ratio between the ground truth path (GT) and the matched path (MP):

$$Accuracy = \frac{1}{M}\sum_{m=0}^{M-1}\frac{|GT[m]\cap MP[m]|}{|GT[m]\cup MP[m]|}. \tag{12}$$

Table 2 presents a comparison of the accuracy of the selected map matching algorithms by the described criteria. The "Not matched" column contains the number of trajectories, for which the map matching procedure was not completed.

Table 2. Accuracy measurements of map matching algorithms

Map matching algorithm	Samara dataset			Public dataset		
	RMF	Accuracy	Not matched	RMF	Accuracy	Not matched
DBA base	0.135	0.876	0	0.286	0.714	7
DBA modified	**0.04**	**0.96**	0	**0.059**	**0.925**	0
FMM	0.245	0.836	5	0.079	0.919	9
HMM	0.744	0.433	0	2.65	0.24	3

The accuracy of the proposed modified algorithm is higher than the accuracy of the baseline map matching algorithms and the base version of the proposed algorithm on both datasets. It also should be noted that only a modification of the proposed map matching algorithm was able to process all the trajectories from the two datasets.

To visually evaluate the quality of the algorithms, ground truth paths and matched paths were displayed on the map. Figure 3 shows examples of the map matching result. The ground truth path is shown by the dash black line, the corresponding map matching algorithm result is shown by the green line. As can be seen, the modification of the proposed algorithm allows to significantly increase the quality of the map matching in comparison with the original algorithm. In comparison with the FMM algorithm result, the modified version of the proposed algorithm shows slightly higher accuracy.

5 Conclusions

Accurate map matching is a challenging problem that has drawn great research attention in recent years. Map matching algorithms aim to reduce the uncertainty in a trajectory and identify the correct path by matching the GPS points to the road network on a digital map. The correction of the raw positioning data has been important for many applications such as navigation and tracking systems.

This work contributes to the existing research on the map matching problem. In particular, we consider the offline map matching problem in which the whole trajectory is processed after it has been collected. The new algorithm based on a dynamic programming approach has been developed. The proposed algorithm consists of two steps performed in a cycle: path estimation and sequential removal of edges from the path. To estimate the matching path, we presented a map

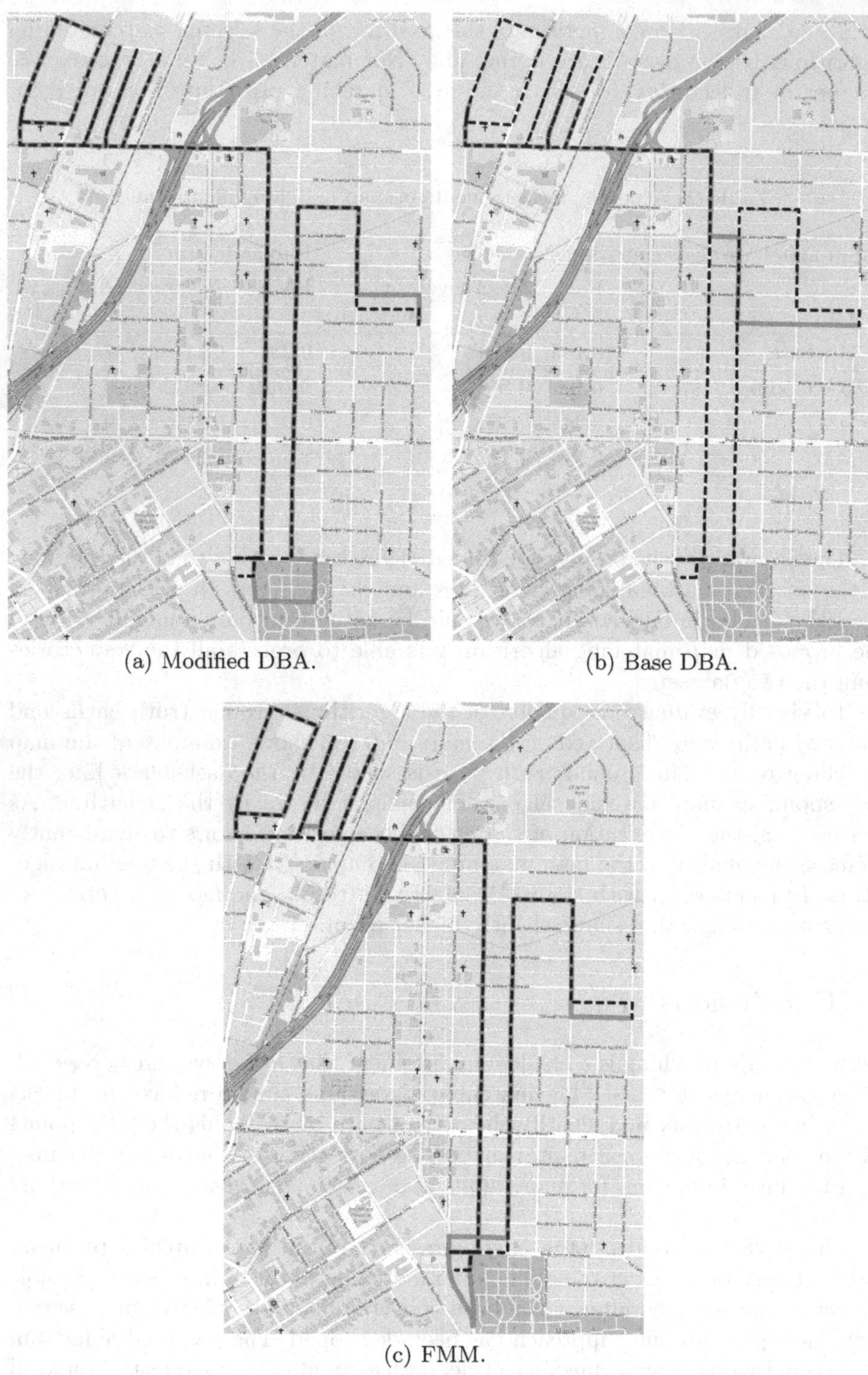

(a) Modified DBA.

(b) Base DBA.

(c) FMM.

Fig. 3. Examples of the map matching result. (Color figure online)

matching algorithm based on a dynamic programming approach. We propose a modification of the proposed map matching algorithm aimed at improving the accuracy of the map matching. The modification of the algorithms consists in dividing the GPS trajectory into sections and step-by-step running the base map matching algorithm for each section.

Experiments were conducted on two datasets. The first dataset is based on a large-scale transportation network of Samara. The second dataset is a publicly available dataset that contains trajectories from around the world. Experiments demonstrated that the proposed algorithm has high accuracy and outperforms other comparable methods by selected criteria. In particular, the proposed algorithm showed slightly better results in terms of accuracy on a publicly available dataset and significantly superior baseline algorithms on a Samara dataset. Moreover, the proposed algorithm made it possible to match all the considered tracks.

The research findings can be used by scholars and practitioners to improve the quality of a range of intelligent transport system applications and services such as route guidance, fleet management, road user charging, accident and emergency response, bus arrival information, and other location based services that require location information.

In this paper, we focused on developing a map matching algorithm that can match GPS trajectories with high accuracy. As a result, the main limitation of the proposed algorithms is the high running time, which does not allow using the proposed algorithm for solving the map matching problem in real time. In future studies, we plan to improve the performance of the proposed algorithm by introducing precomputation techniques and parallel computing.

References

1. Agafonov, A., Yumaganov, A.: Short-term traffic flow forecasting using a distributed spatial-temporal k nearest neighbors model. In: Proceedings of the 21st IEEE International Conference on Computational Science and Engineering, CSE 2018, pp. 91–98 (2018). https://doi.org/10.1109/CSE.2018.00019
2. Nagy, A., Simon, V.: Identifying hidden influences of traffic incidents' effect in smart cities. In: Proceedings of the 2018 Federated Conference on Computer Science and Information Systems, FedCSIS 2018, pp. 651–658 (2018). https://doi.org/10.15439/2018F194
3. Amara, Y., Amamra, A., Daheur, Y., Saichi, L.: A GIS data realistic road generation approach for traffic simulation. In: Proceedings of the 2019 Federated Conference on Computer Science and Information Systems, FedCSIS 2019, pp. 385–390 (2019). https://doi.org/10.15439/2019F223
4. Agafonov, A., Myasnikov, V., Borodinov, A.: Anticipatory vehicle routing in stochastic networks using multi-agent system. In: Proceedings of the 2019 21st International Conference on "Complex Systems: Control and Modeling Problems", CSCMP 2019, vol. 2019-September, pp. 91–95 (2019). https://doi.org/10.1109/CSCMP45713.2019.8976557
5. Carchiolo, V., Loria, M.P., Malgeri, M., Modica, P.W., Toja, M.: An adaptive algorithm for geofencing. In: Ziemba, E. (ed.) AITM/ISM 2018. LNBIP, vol. 346, pp. 115–135. Springer, Cham (2019). https://doi.org/10.1007/978-3-030-15154-6_7

6. Feng, F., Pang, Y., Lodewijks, G.: Towards context-aware supervision for logistics asset management: concept design and system implementation. In: Ziemba, E. (ed.) AITM/ISM 2016. LNBIP, vol. 277, pp. 3–19. Springer, Cham (2017). https://doi.org/10.1007/978-3-319-53076-5_1
7. Da Silva, D., et al.: Inference of driver behavior using correlated IoT data from the vehicle telemetry and the driver mobile phone. In: Proceedings of the 2019 Federated Conference on Computer Science and Information Systems, FedCSIS 2019, pp. 487–491 (2019). https://doi.org/10.15439/2019F263
8. Myasnikov, V.: Reconstruction of functions and digital images using sign representations. Comput. Opt. **43**(6), 1041–1052 (2019). https://doi.org/10.18287/2412-6179-2019-43-6-1041-1052
9. Kubička, M., Cela, A., Mounier, H., Niculescu, S.: On designing robust real-time map-matching algorithms. In: 2014 17th IEEE International Conference on Intelligent Transportation Systems, ITSC 2014, pp. 464–470 (2014). https://doi.org/10.1109/ITSC.2014.6957733
10. White, C., Bernstein, D., Kornhauser, A.: Some map matching algorithms for personal navigation assistants. Transp. Res. Part C: Emerg. Technol. **8**(1–6), 91–108 (2000). https://doi.org/10.1016/S0968-090X(00)00026-7
11. Wei, H., Wang, Y., Forman, G., Zhu, Y., Guan, H.: Fast Viterbi map matching with tunable weight functions. In: Proceedings of the 20th International Conference on Advances in Geographic Information Systems, pp. 613–616. ACM, New York (2012). https://doi.org/10.1145/2424321.2424430
12. Fiedler, D., Čáp, M., Nykl, J., Žilecký, P., Schaefer, M.: Map matching algorithm for large-scale datasets. arXiv:1910.05312 [cs, eess] (2019)
13. Li, Y., Huang, Q., Kerber, M., Zhang, L., Guibas, L.: Large-scale joint map matching of GPS traces. In: Proceedings of the 21st ACM SIGSPATIAL International Conference on Advances in Geographic Information Systems, pp. 214–223. ACM, New York (2013). https://doi.org/10.1145/2525314.2525333
14. Kubicka, M., Cela, A., Mounier, H., Niculescu, S.I.: Comparative study and application-oriented classification of vehicular map-matching methods. IEEE Intell. Transp. Syst. Mag. **10**(2), 150–166 (2018). https://doi.org/10.1109/MITS.2018.2806630
15. Hashemi, M., Karimi, H.A.: A critical review of real-time map-matching algorithms: current issues and future directions. Comput. Environ. Urban Syst. **48**, 153–165 (2014). https://doi.org/10.1016/j.compenvurbsys.2014.07.009
16. Chao, P., Xu, Y., Hua, W., Zhou, X.: A survey on map-matching algorithms. In: Borovica-Gajic, R., Qi, J., Wang, W. (eds.) ADC 2020. LNCS, vol. 12008, pp. 121–133. Springer, Cham (2020). https://doi.org/10.1007/978-3-030-39469-1_10
17. Karimi, H.A., Conahan, T., Roongpiboonsopit, D.: A methodology for predicting performances of map-matching algorithms. In: Carswell, J.D., Tezuka, T. (eds.) W2GIS 2006. LNCS, vol. 4295, pp. 202–213. Springer, Heidelberg (2006). https://doi.org/10.1007/11935148_19
18. Quddus, M., Ochieng, W., Noland, R.: Current map-matching algorithms for transport applications: state-of-the art and future research directions. Transp. Res. Part C: Emerg. Technol. **15**(5), 312–328 (2007). https://doi.org/10.1016/j.trc.2007.05.002
19. Jagadeesh, G., Srikanthan, T., Zhang, X.: A map matching method for GPS based real-time vehicle location. J. Navig. **57**(3), 429–440 (2004). https://doi.org/10.1017/S0373463304002905

20. Joshi, R.: A new approach to map matching for in-vehicle navigation systems: the rotational variation metric. In: IEEE Conference on Intelligent Transportation Systems, Proceedings, ITSC, pp. 33–38 (2001)
21. Srinivasan, D., Cheu, R., Tan, C.: Development of an improved ERP system using GPS and AI techniques. In: Proceedings of the IEEE Conference on Intelligent Transportation Systems, ITSC, pp. 554–559 (2003). https://doi.org/10.1109/ITSC.2003.1252014
22. Brakatsoulas, S., Pfoser, D., Salas, R., Wenk, C.: On map-matching vehicle tracking data. In: VLDB 2005 - Proceedings of 31st International Conference on Very Large Data Bases, pp. 853–864 (2005)
23. Yin, H., Wolfson, O.: A weight-based map matching method in moving objects databases. In: Proceedings of the International Conference on Scientific and Statistical Database Management, SSDBM, vol. 16, pp. 437–438 (2004)
24. Li, L., Quddus, M., Zhao, L.: High accuracy tightly-coupled integrity monitoring algorithm for map-matching. Transp. Res. Part C: Emerg. Technol. **36**, 13–26 (2013). https://doi.org/10.1016/j.trc.2013.07.009
25. Velaga, N.R., Quddus, M.A., Bristow, A.L.: Developing an enhanced weight-based topological map-matching algorithm for intelligent transport systems. Transp. Res. Part C: Emerg. Technol. **17**(6), 672–683 (2009). https://doi.org/10.1016/j.trc.2009.05.008
26. Sharath, M., Velaga, N., Quddus, M.: A dynamic two-dimensional (D2D) weight-based map-matching algorithm. Transp. Res. Part C: Emerg. Technol. **98**, 409–432 (2019). https://doi.org/10.1016/j.trc.2018.12.009
27. Chen, B., Yuan, H., Li, Q., Lam, W., Shaw, S.L., Yan, K.: Map-matching algorithm for large-scale low-frequency floating car data. Int. J. Geogr. Inf. Sci. **28**(1), 22–38 (2014). https://doi.org/10.1080/13658816.2013.816427
28. Syed, S., Cannon, M.: Fuzzy logic based-map matching algorithm for vehicle navigation system in Urban Canyons. In: Proceedings of the National Technical Meeting, Institute of Navigation, pp. 982–993 (2004)
29. Quddus, M., Noland, R., Ochieng, W.: A high accuracy fuzzy logic based map matching algorithm for road transport. J. Intell. Transp. Syst.: Technol. Plann. Oper. **10**(3), 103–115 (2006). https://doi.org/10.1080/15472450600793560
30. Newson, P., Krumm, J.: Hidden Markov map matching through noise and sparseness. In: GIS: Proceedings of the ACM International Symposium on Advances in Geographic Information Systems, pp. 336–343 (2009). https://doi.org/10.1145/1653771.1653818
31. Raymond, R., Morimura, T., Osogami, T., Hirosue, N.: Map matching with Hidden Markov Model on sampled road network. In: Proceedings of the International Conference on Pattern Recognition, pp. 2242–2245 (2012)
32. Yin, Y., Shah, R., Zimmermann, R.: A general feature-based map matching framework with trajectory simplification. In: Proceedings of the 7th ACM SIGSPATIAL International Workshop on GeoStreaming, pp. 1–10. ACM, New York (2016) https://doi.org/10.1145/3003421.3003426
33. Yang, C., Gidófalvi, G.: Fast map matching, an algorithm integrating hidden Markov model with precomputation. Int. J. Geogr. Inf. Sci. **32**(3), 547–570 (2018). https://doi.org/10.1080/13658816.2017.1400548
34. de Sousa, R.S., Boukerche, A., Loureiro, A.A.F.: A map matching based framework to reconstruct vehicular trajectories from GPS datasets. In: 2020 IEEE International Conference on Communications (ICC), ICC 2020, pp. 1–6 (2020). https://doi.org/10.1109/ICC40277.2020.9148732

35. Yumaganov, A., Agafonov, A., Myasnikov, V.: Map matching algorithm based on dynamic programming approach. In: 2020 15th Conference on Computer Science and Information Systems (FedCSIS), pp. 563–566 (2020). https://doi.org/10.15439/2020F139
36. Kubicka, M., Cela, A., Moulin, P., Mounier, H., Niculescu, S.: Dataset for testing and training of map-matching algorithms. In: Proceedings of the IEEE Intelligent Vehicles Symposium, pp. 1088–1093 (2015). https://doi.org/10.1109/IVS.2015.7225829
37. GraphHopper library (2020). https://github.com/graphhopper/map-matching

A Cluster-Based Approach to Solve Rich Vehicle Routing Problems

Emir Zunic[1,2(✉)], Sead Delalic[1,3], Dzenana Donko[2], and Haris Supic[2]

[1] Info Studio d.o.o., Sarajevo, Bosnia and Herzegovina
emir.zunic@infostudio.ba
[2] Faculty of Electrical Engineering, University of Sarajevo, Sarajevo, Bosnia and Herzegovina
{ddonko,hsupic}@etf.unsa.ba
[3] Faculty of Science, University of Sarajevo, Sarajevo, Bosnia and Herzegovina
delalic.sead@pmf.unsa.ba

Abstract. The vehicle routing problem is one of the most difficult combinatorial optimization problems. Due to a large number of possibilities and practical limitations, there is no known algorithm that finds the optimal solution. Heuristics and metaheuristics have shown quality results in solving routing problems, however, for large instances of the problem these approaches have also shown weaknesses. The paper presents a two-phase algorithm to solve rich vehicle routing problems. The approach is based on customer clustering using a discrete Firefly algorithm, and solving an individual vehicle routing problem for each created cluster, with methods of sharing information and resources among clusters. Due to the smaller dimension, routing problems within a cluster can be solved more easily. The approach has been tested in practice in the creation of routes for large warehouses and has given quality results. Savings of between 15–20% were achieved during more than 30 days of testing. The obtained routes do not violate practical restrictions and can be used in practice. The increase in the number of customers does not significantly affect the complexity of the solution, which is a great advantage over standard methods.

Keywords: Vehicle routing problem · Rich vehicle routing problem · Firefly algorithm · Clustering · Combinatorial optimization

1 Introduction

The problem of transport management is an important group of processes whose optimization can achieve significant savings for the company. Transport is an important factor in the work of distribution and other companies with the delivery and transport of goods.

Delivery of goods to customers should be the end result of a series of processes of preparation of goods, transport planning and transport to the customer. Due to the company's reputation, but also the large number of customers and the

E. Ziemba and W. Chmielarz (Eds.): ISM 2020/FedCSIS-IST 2020, LNBIP 413, pp. 103–123, 2021.
https://doi.org/10.1007/978-3-030-71846-6_6

reduction of costs, it is of great importance that the transport is an efficient system.

Transport management system is composed of several elements important for planning, execution, optimization and monitoring of goods transport. Route planning is the most important factor. This problem has been the focus of a large number of research since the first mention in the 1950s [1].

The Vehicle Routing Problem (VRP) is a problem of determining the delivery vehicle for each customer, and the order of customers to be delivered for each vehicle in the fleet. The ultimate goal is to minimize the cost. Routing problem is made up of a number of necessary elements, such as the vehicle fleet and customers with logistic data, the number of drivers and workers loading goods, real-world limitations and more. By transport optimization, significant savings are achieved and the efficiency of the company is increased.

A number of different versions of the problem are known. For a successful routing, it is necessary to provide a number of logistical data on customers and the vehicle fleet. Through years of Vehicle Routing Problem research, a number of versions have been described, and most are based on an analysis of realistic constraints.

Customers set a time window when the orders can be delivered, in which case the Vehicle Routing Problem with Time Windows (VRPTW) is discussed. Each vehicle has a limited capacity, and the Capacited Vehicle Routing Problem (CVRP) is often described in the literature, where the volume and mass of goods in the vehicle cannot be violated. In addition, the vehicle fleet is often heterogeneous, which further complicates the process of finding a solution. In a heterogeneous fleet of vehicles, there are limitations where a particular vehicle cannot deliver goods to a particular customer (e.g., a truck cannot deliver goods to the city center).

Problems that involve a number of different constraints are called Rich Vehicle Routing Problems (RVRP). By observing realistic constraints, the number of possibilities is significantly reduced, and finding a feasible solution is an important step in optimization.

The number of customers depends on the size of the company. In practice, there are frequent situations where it is necessary to route hundreds of customers, which significantly increases the number of options and slows down the routing process. Therefore, customer clustering is often approached, as well as solving a minor vehicle routing problem for each cluster.

Customers are most often divided into clusters based on geographical location, by city or region. In doing so, it is necessary to determine a city, region or cluster for each customer, which often slows down the routing process. This method of division leads to several additional problems in regions with dense customers. The paper describes the approach to automatic solution of clustering problems with the aim of automating the complete routing process. The complete approach is based on many years of experience in the work of distribution companies and routing vehicles. An innovative version of the discrete Firefly algorithm (DFA) has been implemented to create the clusters.

The paper consists of five sections. In the first section, the RVRP is defined and the motive for solving the routing problem by applying clustering methods is described. In the second section, an extensive overview of the literature is given. The third section describes a proposed two-phase method for solving routing problems based on a discrete Firefly algorithm and solving minor VRP problems. In the fourth section, test results for real-world distribution company are given. In the last section, a conclusion is described and guidelines for future work are given.

2 Literature Review

The problem of vehicle routing belongs to a class of hard optimization problems. The problem-solving algorithm is not known, and the only method is space search. The solution space contains a large number of possibilities, and checking each possibility does not lead to success. In this case, heuristic approaches are applied. A special group of algorithms that give notable results in solving hard optimization problems are nature-inspired metaheuristic algorithms.

A large number of metaheuristic algorithms from the group of swarm intelligence (SI) algorithms are known, such as the Particle Swarm Optimization (PSO), Bat Algorithm (BA), Firefly Algorithm (FA), Cuckoo search (CS), Flower Pollination Algorithm (FPA) described in [2], or Fireworks Algorithm (FWA) [3], Ant Colony Optimization (ACO) [4] and Elephant Herding Optimization (EHO) [5].

Nature-inspired metaheuristic algorithms have important applications in solving optimization problems in the field of Supply Chain Management (SCM), logistics and engineering. In the paper [6], the concept of smart Warehouse Management System (WMS) is described. The concept has been improved by applying the ACO algorithm for splitting [7], and the BAT algorithm for batching orders during the picking process [8]. Many other factors can be used in combination with metaheuristic techniques and data collected to improve software systems [9,10]. The efficient application of metaheuristic algorithms to engineering problems is described in [11].

An important warehouse process is the order picking optimization, which is a modification of the Traveling Salesman Problem (TSP). In the paper [12], the process of distance calculation in the warehouse of generic layout is described. TSP is one of the most researched problems in the field of combinatorial optimization, and many metaheuristic algorithms have been applied to problem variants, such as BA [13,14], FA [15], FWA [16], EHO [17], PSO [18] and others.

The problem of vehicle routing is often seen as a generalization of the traveling salesman problem. It represents one of the most important problems in SCM due to the large savings that optimization brings. Although the problem has been in the literature for more than 50 years, the field has been extremely active in recent years. Well-known approaches such as Genetic Algorithm [19], Simulated Annealing [20] or Tabu Search [21] have yielded successful results in the field. A large number of variants of VRP problems are known to arise primarily due to practical limitations.

A large number of approaches to solving the problem of RVRP are known, as well as a large number of variants of the problem [22,23].

The particle swarm optimization is one of the most well-known metaheuristic algorithms, and is considered a pioneering algorithm in the field. Therefore, there are a number of applications of this algorithm for vehicle routing. In the paper [24], PSO algorithm is used to solve the VRP with simultaneous pickup and deliveries. The application of a hybrid algorithm to CVRP resolution is described in [25], as well as to VRPTW in [26] and VRPTW with a heterogeneous vehicle fleet in [27].

Bat algorithm is implemented for different versions of the VRP. In the paper [28], the CVRP problem is solved with adapted BA. In the [29], the hybrid BA with path relinking for the CVRP is presented. An efficient BA with random reinsertation operators is presented for VRPTW in [30] and the chaotic discrete bat algorithm for the CVRP is described in [31].

Ant colony optimization and improved ACO for the VRP are described in [32] and [33]. An artificial bee colony algorithm is implemented for the CVRP [34] and for the periodic vehicle routing problem [35]. Grey wolf algorithm to solve the CVRP is described in [36]. The same problem is solved with the improved fireworks algorithm in [37].

The Firefly algorithm has been implemented in several versions for routing problems. In [38], the authors presented a discrete firefly algorithm for solving RVRP for the distribution of newspapers with recycling policies. An improved hybrid firefly algorithm for CVRP is described in [39]. The VRP with time windows solver based on the firefly algorithm with novel operators is presented in [40]. Firefly algorithm is also implemented for the heterogeneous fixed fleet vehicle routing problem [41].

The problem of vehicle routing for large instances is often solved by dividing customers into clusters, and by individual routing for each cluster. After division into clusters, there is an additional set of unresolved issues, such as the distribution of vehicles by clusters, cluster priorities, the possibility of combining clusters or adding customers to the created routes within the cluster.

There are a number of cluster-based versions of the problem in the literature. In [42], the hybrid metaheuristic algorithms were proposed for the clustered vehicle routing problem. Customers are grouped into clusters, and the vehicle after entering a particular cluster must deliver the goods to all customers in the cluster. An Iterated Local Search algorithm and Hybrid Genetic Search algorithms have been implemented, so as other similar approaches [43,44].

In [45], the two-phase cluster-based approach for solving RVRP is described. Customers are divided into clusters and in the second phase, the smaller VRP is solved for each cluster. Customers are divided into clusters manually or by using k-means [46], k-medoids [47] or hierarchical techniques of clustering [48].

Clustering is a well-known problem in the field of data mining. In addition to the previously mentioned methods, there are a number of algorithms for data clustering, with special emphasis on the domain of the problem. A special group consists of the application of metaheuristic algorithms to solve clustering problems. In [49], the applications of the Firefly algorithm to solve clustering problems are described. An improved k-means algorithm based on the Firefly algorithm has been implemented [50]. Some other nature-inspired metaheuristic clustering algorithms have been implemented, such as PSO [51], genetic algorithm [52], simulated annealing [53], and others.

For distribution companies, customers are often repetitive, therefore, there are a number of opportunities to improve the input for future optimization. Vehicles are often tracked through a GPS system, and their movements are fully recorded and available for analysis. To apply data analysis algorithms, it is necessary to ensure the correctness of the data [54]. A number of improvements to vehicle routing algorithms based on GPS data analysis have been described in [55]. Unloading time is one of the most important factors of RVRP. The application of machine learning techniques and GPS data to improve the unloading time prediction for VRP is described in [56]. The complete concept of innovative comparison method for route evaluation is described in [57].

3 Case Study

The problem of solving vehicle routing problem has been reduced to two steps: dividing customers to clusters, and solving a vehicle routing problem for each cluster. In the first phase, a number of clusters were determined based on the size of the vehicle fleet and the number of customers, as well as other logistical data. For the clustering process, a discrete Firefly algorithm was applied. In the second phase, a modified 2-opt algorithm and Tabu search were used to solve a number of minor routing problems. The process is shown in Fig. 1.

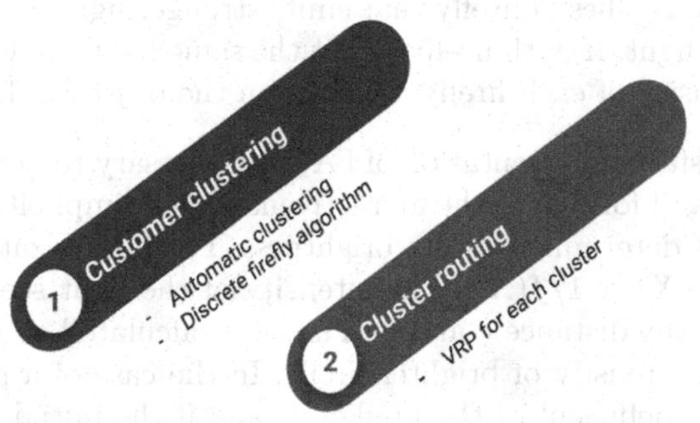

Fig. 1. Two-phase approach for the vehicle routing problem

The exact location (latitude and longitude) was determined for the depot, as well as data on all available vehicles and customers. For each vehicle, data on maximum capacity, cost per kilometer and operating hours were collected. For each customer, the exact location (latitude and longitude), data on ordered items (volume, mass, number of items), estimated unloading time, and time window are determined. Data on all customer-vehicle constraints were also collected and included in the cost calculation process. All distances between customers as well as distances between depot and customers were calculated. The unloading time was estimated based on the formula and machine learning techniques described in [56]. The goal of the proposed algorithm is to deliver orders to all customers, and to minimize the total cost.

3.1 Customer Clustering

The problem of automatic clustering was solved using a nature-inspired metaheuristic Firefly algorithm (FA). The Firefly algorithm was created by Xin-She Yang at the Cambridge University to solve continuous optimization problems. The algorithm is inspired by the light displayed by the fireflies during the summer or in tropical regions. The main purpose of light production is to attract female partners, attract prey, and warn opponents [2].

The intensity of the light emitted by the fireflies affects the attraction. At the same time, the distance between the fireflies significantly affects the intensity of the light that the other firefly sees. Therefore, the light intensity, denoted by I, decreases with increasing distance r between the two fireflies. The intensity of light is most often associated with the objective function, which allows the creation of a new nature-inspired metahreuristic algorithm.

As described in [58], the Firefly algorithm uses three simplified rules of firefly behavior:

1. Fireflies are unisex. Each firefly will attract and be attracted to all other fireflies.
2. The attractiveness of the fireflies is proportional to the intensity of their light. For each pair of fireflies, a firefly that emits stronger light will attract a firefly that emits less light. If both fireflies emit the same light, they move randomly.
3. The light intensity of each firefly depends on the objective function.

For the successful implementation of FA it is necessary to define the variation of light intensity and formulate the attractiveness. For simplicity, the attractiveness of a firefly is determined by its brightness. For minimization problems, it is assumed that $I(X) \propto 1/f(X)$. The intensity of the light seen by the weaker firefly depends on the distance r and it is usually calculated as $I(r) = \frac{I_s}{r^2}$, where the I_s is the light intensity of brighter firefly. In the case of a previously determined absorption coefficient of the fireflies γ, and if the initial light intensity is I_0, the value of I is calculated as

$$I = I_0 \cdot e^{-\gamma r^2}. \tag{1}$$

As the attractiveness of the fireflies is proportional to the intensity of the light, the attractiveness β is defined as

$$\beta = \beta_0 e^{-\gamma r^2}, \tag{2}$$

where β_0 is attractiveness at the distance $r = 0$. To further speed up the algorithm, the exponential function is changed by the function $1/(1 + r^2)$, and the attractiveness is expressed by the relation

$$\beta = \frac{\beta_0}{1 + \gamma r^2}. \tag{3}$$

The motion of each firefly is defined by the

$$x_i = x_i + \beta \cdot (x_j - x_i) + \alpha \epsilon_i, \tag{4}$$

where the firefly x_i is attracted by firefly x_j, the β is calculated by formula 2 or 3 with the distance r_{ij}. Alpha is a randomization parameter, and ϵ_i is a vector of random numbers obtained from a Gaussian or uniform distribution.

Algorithm 1. Firefly Algorithm For Clustering

```
Find the number of clusters k.
Define the objective function f(X) where X = (x_1, x_2, ... x_k).
Generate the initial population of n fireflies x_i, where i = 1,n (overlined).
Formulate the light intensity I as a function of f(X).
Define the light absorption coefficient γ.
Define the maximum generation number MaxGeneration and the counter t.
while t < MaxGeneration do
    for i = 1 to n do
        for j = 1 to n do
            if I_i < I_j then
                Move firefly i towards j
            end if
            Vary attractiveness with distance r via exp −γr
            Evaluate new solutions and update light intensity
        end for
    end for
    Rank fireflies and find the current global best g*
end while
return g*
```

The described algorithm is intended to solve the problem of continuous optimization. The problem of clustering can be considered as a problem of continuous optimization, as described in the paper [59]. In doing so, for each cluster the centroid is determined as a free location in space (pair of latitude and longitude). The paper will describe the approach to solving the problem where the centroids of the cluster are observed in relation to customer locations. Therefore, the problem reduces to the discrete version, and the original FA must be adapted. In this way, the search space is significantly reduced.

The pseudocode for the discrete FA is given in Algorithm 1. To implement a discrete version of the algorithm, it is necessary to customize several elements. It is necessary to calculate the number of clusters, to create the initial population and to determine the objective function. As the main part of the algorithm, it is necessary to define the movement operator for each firefly.

Number of Clusters. The initial step in the process of dividing customers into clusters is to determine the number of necessary clusters. The number of clusters is denoted by k. Let c denote the total number of customers, and let v denote the total number of vehicles. Then the number of clusters is calculated using the equation

$$k = \lceil \frac{c}{v \cdot 2} \rceil + 1. \quad (5)$$

The basic idea is to use 2 vehicles per sector. The number of vehicles per cluster may vary, but smaller clusters guarantee faster solutions and exploration of fewer options during routing within a cluster.

Initial Population. In order to create an initial population composed of n fireflies, it is necessary to determine k locations for each individual. Each location will represent the center of the cluster (centroid). For each customer, the cluster is determined based on the nearest centroid. So, of all the proposed centroids, the closest one is chosen, and the customer is added to the appropriate cluster.

For the discrete version of the problem, the centroids are selected from the original customer list. So, from the original list of c customers, a randomly selected combination of k customers is selected, and the process is repeated. For each individual, the corresponding clusters for each customer are determined, as well as the final value of the objective function and light intensity.

Objective Function. The biggest advantage of the Firefly algorithm and meta-heuristic approaches over classical clustering methods is the ability to change the objective function depending on the needs of the company, and obtain quality results even after the modification.

Initially, the objective function is calculated in a standard way, where the corresponding cluster is determined for each customer, and the total sum of the distance of all customers from the central location of the corresponding clusters is returned as the value of the objective function. The arithmetic mean of the positions of all customers in the cluster is taken as the central location. The ultimate goal is certainly to minimize the value obtained.

A number of modifications of the goal function were also implemented, where the previously obtained value was taken as one of the factors of the goal function. Other factors are the balance in the number of customers in each cluster (where the number of customers within the cluster tends to be balanced) or the number of customers with restrictions in each cluster.

The Movement Operator. The movement operator for each firefly x is determined based on its current position and the position of the other firefly y. In doing so, without losing generality, it can be assumed that the firefly y is in a better position, and that it has a smaller value of the objective function.

Each firefly is determined by an array of customers who represent centroids. Each array is sorted by the original array of customers. Thus, if the centroids c_i and c_j are chosen in an individual x, and if $i < j$, then in the array of centroids c_i is before c_j. Let the centroid be represented by the customer positions in the original array of customers. Therefore, in the previous case, firefly x would contain the values i and j.

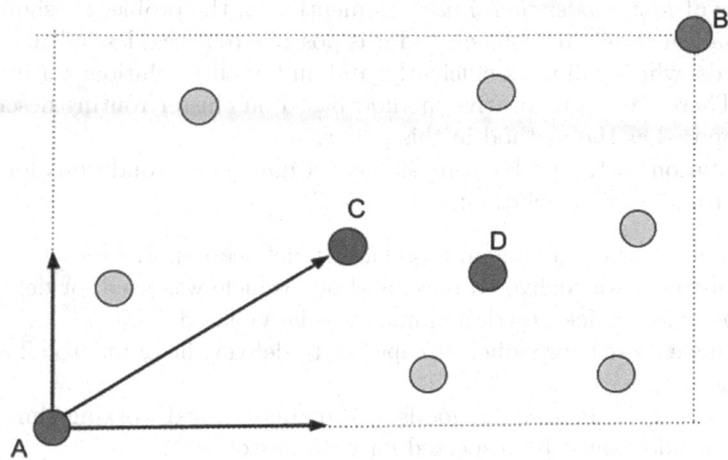

Fig. 2. Movement of the firefly

For each centroid x_i, the corresponding centroid y_i was found, and the distances by latitude and longitude were calculated. Each of the distances is multiplied by a random real number between 0 and 2 and added to the old latitude and longitude values by the Eq. 4. The obtained values are marked as the new potential position of the centroid. The actual position of the centroid is calculated as the position of the nearest customer from the original list to the potential position. The process is shown in Fig. 2. A represents the original location of the observed centroid moving towards the original location of the corresponding centroid B in the better firefly. The initial result of the move is location C. However, of all the other customers, the one closest to location C is selected, and this customer is selected as the centroid (location D in the figure).

Other Elements. The implementation of other elements does not differ from the proposed method in [2].

3.2 Cluster Routing

After the successful division of customers into clusters, the second phase of the algorithm is approached. The second phase comes down to solving the vehicle routing problem for each individual clusters. In doing so, it is important to note that the problem is significantly smaller in size than the original problem. However, it is also important to emphasize that when routing, it is necessary to ensure the exchange of data between clusters, and to provide several additional improvements to the complete process. Due to the size of the problem and the number of minor VRPs to resolve, it is advisable to use a simple and fast approach to find routes, such as the 2-opt algorithm.

One of the methods used is described in the paper [45], where after clustering the approach to solving the problem of routing in several steps is based on the philosophy of *first cluster then route*. As mentioned, the problem is significantly smaller than the original problem, and it is possible to access less advanced routing methods, which will work efficiently, and find quality solutions for individual clusters. Therefore, an innovative method based on cluster routing described in [59] is proposed as the method in this paper.

The solution to the problem must meet a number of conditions for finding an efficient and feasible solution:

- Each route starts and ends in a previously defined depot;
- Each customer was delivered once, and one vehicle was used for delivery;
- The customer-vehicle restriction must not be violated;
- The vehicle must have sufficient capacity to deliver the orders to all assigned customers;
- The vehicle must deliver the goods at a predetermined working time;
- A time window must be respected for each customer.

In the event that some of the previous conditions are violated, a penalty value is introduced which increases the goal function, and indicates a reduction in the quality of the proposed solution.

In the first step, cluster D to which the depot belongs is determined. All customers from the specified cluster can be used to supplement routes from other clusters. In case the obtained route for an vehicle is underused, one or more customers from D are selected to be added to the route. This increases the utilization of all vehicles. In case some customers from D remain undelivered, depot D routing is done at the end.

For each of the remaining clusters, a priority is set. Priority can be determined using different approaches. Clusters with higher priority are routed first. In the described approach, the number of customers and the number of restrictions are the criteria for determining the priority. Therefore, clusters with more customers and more customer-vehicle constraints have a higher priority for routing than clusters with fewer customers and clusters with fewer constraints.

The pseudocode for the described approach is given in Algorithm 2. As can be seen, for each cluster, one problem of routing smaller vehicles is solved. For successful implementation, it is necessary to define the goal function, define the VRP problem solving method for a predefined set of vehicles, and define the procedure for adding customers to previously created routes.

Algorithm 2. Cluster routing algorithm [59]

```
Initialize parameters n, α, β, γ.
Define a priority for each cluster CL_i, i = 1..k.
Sort clusters by priority (higher priority before).
Define the objective function f_GOAL.
for CL_i in CL do
    Get all customers C in sector CL_i and free vehicles
    Set number of vehicles to use (i = 1)
    while true do
        Create all combinations V of i free vehicles
        for v in V do
            Solve VRP using vehicles v for customers C
            Update best solution if found
        end for
        if best solution feasible then
            For each vehicle, add customers from D when possible
            break
        end if
        i = i + 1
    end while
end for
Try to swap vehicles to reduce cost and return best solution
```

Objective Function. When defining the goal function, it is important to consider several types of costs for each vehicle. For each vehicle, a fixed cost is defined (registration price, regular service, vehicle price, etc.), variable cost per kilometer and the penalty cost for violating the time windows and working hours. The goal function should support the idea of minimizing the total cost of routing.

For each vehicle, it is necessary to calculate the total distance traveled and the cost of the trip. In doing so, the vehicle does not have the same cost per kilometer in the case of maximum utilized capacity and in the case of unladen driving. Therefore, for each kilometer traveled, the vehicle load is checked and the cost of driving is calculated. Any violation of the customer-vehicle restriction is added as a penalty value, and the same is added in case of violation of time windows and delays in return (violation of working hours).

For each restriction violated, the route is declared unfeasible. When introducing penalties, minor violations are less penalized, thus ranking unfeasible solutions by quality (a solution that violated the time window in 5 min is better than a solution that violated the same time window in an hour, and violated the customer-vehicle and time constraint return to the depot at the same time).

The described way of defining the goal function enables easy modification and customization to the needs of the company.

VRP Solver. The key part is solving the problem of vehicle routing for the observed cluster, where the list of customers and used vehicles is known in advance. Due to the way the cluster is created, the problem is smaller, so a large number of vehicles are not used for routing. At the same time, a solver is called for each vehicle combination. Therefore, it is not necessary to use advanced methods to find a solution, it is more important to find a quick suboptimal solution.

In solving, the 2-opt algorithm for Traveling Salesman Problem (TSP) is used. A simple 2-opt algorithm changes every two customers in the route, and tries to achieve savings in the value of the objective function.

In order to take advantage of the method provided for solving TSP in solving VRP problem, each vehicle is added as a customer. For new customers, the location is taken from the depot, while the distance from each customer is calculated as the distance between the depot and the customer. The distances between each of the two new customers are large values.

The route is represented by an organized $(n+k)$-tuple, where n is the number of customers in the cluster, and k is the number of vehicles. The first element of the $(n+k)$-tuple is the first vehicle, and all customers until the next added customer in the tuple belong to the route of the first vehicle. All customers between the second and third added customer belong to the route of the second vehicle, etc. All customers after the last added customer in the tuple belong to the last vehicle. Due to the large distance of added customers, it is guaranteed that the obtained solution will not contain a consecutive added customers, i.e. each vehicle will be used.

In case the capacity of the vehicle in the combination is less than the necessary capacity for delivering all customers, the combination is not observed, which reduces the number of solving smaller VRPs. In addition, in the case of a large number of vehicles, it is possible to introduce vehicle priorities for each region, and pre-determine a subset of vehicles that can be used for the region (e.g. based on depot distance from centroids or by any predetermined criteria). Thus, the number of observed combinations can be arbitrarily reduced, without at the same time disrupting the process of creating feasible solutions.

Adding Customers to Created Routes. As mentioned earlier, there are frequent situations when the vehicle capacity is not fully utilized when creating routes for individual clusters. At the same time, the vehicle was returned to the depot earlier, and at the same time it could be used to deliver additional customers. In this case, the routes are supplemented by customers from the previously defined cluster D. Of all customers from cluster D who have not been previously visited, the customer closest to the initial and final customer within the route other than the depot is determined. Let the nearest customer be determined with A. Customer A is added to the beginning or end of the route, and the cost obtained is checked. If no restrictions are violated, the customer remains on the route, and the procedure is repeated as much as possible.

During the implementation, a variant of adding customers to created routes was made, in which the main goal is route balancing. In this case, customers are not added immediately after creating the route, but after routing all clusters except D. For each vehicle, the priority is determined. The highest priority has the vehicle that is the least loaded (it returns to the depot the earliest or has the most free capacity). After adding one customer to that vehicle as described above, the procedure is repeated, and the vehicle with the highest priority is selected again.

Improvements. The concept has been improved by adding three steps. The steps are based on experience in vehicle routing for distribution companies in order to address the identified shortcomings and to improve the previously described process:

- After successfully completing the algorithm and routing of all individual clusters, and finding a solution, an additional check was made in which every two vehicles were changed. After each change, the cost is calculated. If the cost is reduced, the vehicles remain changed. This step can bring savings due to underutilized vehicle capacity or driving costs;
- Occasionally, two situations occur during clustering. In the first case, clusters were created with a smaller number of customers, while in the second case, the clusters are on the same path from the depot. We say that clusters are on the same path, if their centroids are almost collinear points with the depot. We say that two points are almost collinear if a triangle of sufficiently small area is created. In this case, the algorithm allows regions to be merged, creating a larger routing region;
- The third improvement occurs if the vehicle, regardless of balancing and adding customers from cluster D to the route, returns to the initial depot earlier. In that case, it is necessary to use the vehicle for the following routes. If the vehicle is returned to the depot earlier, the route is saved, and other routes within the cluster are canceled and routing is done again. This reuses the vehicle for the same clusters, but with fewer routing customers.

4 Results Discussion

The described approach was created through several years of experience in the process of routing the delivery of distribution companies. A complete approach has been gradually developed. Initially, routing was done manually. The first algorithmic approach was based on Tabu search, which showed a number of weaknesses for practical use. For routing a large number of customers, the algorithm often provided unstable and not feasible solutions in practice.

In the next phase, the clustering approach was implemented. Customers were manually divided based on geographical location (city or region to which they belong). The method proved to be of good quality for practical use, but shortcomings were noticed in urban regions with a larger number of customers. If a large number of customers are in the same city, large are created clusters and the problem of routing becomes significantly more complex. The second drawback is the impossibility of automating the complete process, because it is necessary to divide the customers before performing the routing step.

An approach has been created in which customers are divided based on standard clustering algorithms, such as k-means or k-NN algorithms. The cluster-based approach is described in the paper [45]. It has significantly improved the obtained results for practical use, but at the same time it has been noticed that solving clustering problems can be further improved, and that the principle of solving smaller problems can be modified. The smaller VRPs can be solved with simpler and more efficient methods. The technique based on the standard Firefly algorithm and modified version of the second phase is described in [59].

This paper presents a modification of the Firefly algorithm used for clustering. The method is based on observing the problem of clustering as a problem of combinatorial optimization. The centroids are selected from the original set of customers and cannot be located in arbitrary locations. The implemented approach was tested on a real data. Excellent results were achieved, where quality and feasible solutions were obtained for all tested instances.

Clustering algorithms were tested on real data collected through 30 days of warehouse operation. Figure 3 shows a comparison of the clustering methods for one day. Methods of manual clustering, k-means algorithm, as well as standard and discrete versions of FA are presented. Discrete FA shows the highest quality results for the observed goal function and test data.

As can be seen in the figure, the manual clustering algorithm placed all customers from the densest region in one cluster, which is a significant problem. The k-means algorithm and the standard FA showed quality results in some cases. However, due to the discretization of the problem and the efficient implementation of the movement operator, DFA searches a significantly smaller space, and therefore finds solutions more efficiently. Discrete FA achieved a better value of the goal function in 55% of cases compared to other approaches. In other cases, no significant difference was observed with respect to the FA and k-means algorithm.

Figure 4 shows the proposed route for one vehicle obtained after the last step of the proposed algorithm. A red marker shows the location of the depot. The red line surrounds the cluster to which the depot belongs. As can be seen, the route contains customers from the cluster to which the depot initially belongs.

More than 30 days of actual warehouse operation were taken into consideration during testing. The obtained results of the comparison of the implemented approaches are shown in Table 1. Tabu search violated restrictions in 10 days. Other approaches did not violate any restrictions. The discrete and standard FA were run 10 times, with the best clustering result observed during the second phase of the algorithm. Standard algorithm settings were used, where $n = 20$, $\beta_0 = 1$, $\gamma = 0.1$, $\alpha_0 = 1$, $maxGeneration = 1000$, and $\theta = 0.98$.

Tabu search gave up to 10 times slower results compared to clustering-based methods. Clustering methods have yielded results applicable in practice in a significantly shorter time. The complete process is accelerated, because the expected input from the user is minimal. Clusters are created automatically and the clustering method is adapted to solve this problem. Manual and k-means clustering

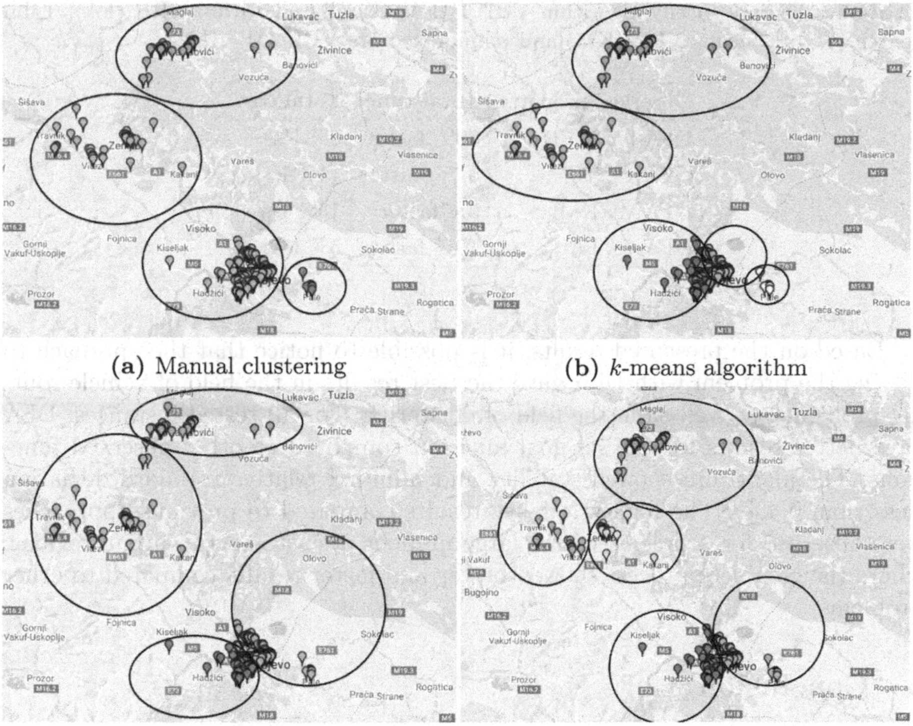

(a) Manual clustering

(b) k-means algorithm

(c) Firefly algorithm

(d) Discrete firefly algorithm

Fig. 3. Clustering algorithms

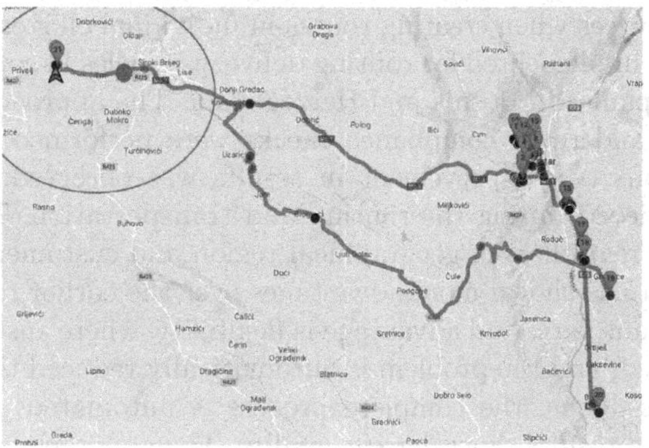

Fig. 4. The proposed route for one vehicle

algorithm are implemented in a combination with Tabu search, because for larger instances the 2-opt algorithm showed a significantly greater possibility of finding a suboptimal solution for the proposed goal function.

Table 1. Discrete Firefly algorithm VRP (DFA), Firefly algorithm VRP (FA), Tabu Search (TS), k-means VRP (KM) and manual sectors VRP (MVRP)

Observed property	Total time	Total cost
DFA vs. MVRP	30% faster	6% less
DFA vs. TS	5.7× faster	11% less
DFA vs. KM	5× faster	15% less
DFA vs. FA	10% faster	5% less

Based on the presented results, it is possible to notice that the approach to solving the problem with DFA gave the best results in the field of vehicle routing problems, as well as in the field of clustering. For all tested instances, DFA gave better results for the set goal function compared to other observed functions. The algorithm showed stability and minimal relative standard deviation (less than 0.4%). The highest quality results compared to previous approaches were observed for problems with a large number of customers in urban regions, where the complete process showed better and faster results compared to other problems.

5 Conclusion

The paper describes an innovative two-phase approach to solving routing problems. The approach was used to address the rich VRP. In the first step, the discrete Firefly algorithm is used to divide customers into clusters. The second phase consists of solving a number of minor VRP problems, and optimizing the allocation of resources when creating routes in individual clusters. The complete concept has been implemented for routing delivery vehicles in some of the largest distribution companies in Bosnia and Herzegovina. The approach was tested on real-world data, and route compliance checks were performed with previously defined restrictions. An improvement in results was observed, and significant savings were achieved during the routing and transportation. The process can be used globally, regardless of geographical region and customer location.

The approach has shown more advantages over the earlier routing methods. The most important perceived advantage is flexibility, where customers are automatically clustered, and the problem is automatically reduced to solving several smaller problems. Thus, the complete process is automated and accelerated, while the obtained results showed great quality. The approach offers a reduction in the complexity of the problem due to the customer clustering process, as well as other improvement techniques, such as unloading time prediction based on GPS data history.

In the future, it is planned to apply other metaheuristic approaches to solving customer clustering problems, such as the bat algorithm or the fireworks algorithm. In addition, it is planned to add a number of practical restrictions when creating clusters based on the analysis of previously obtained and driven routes.

Other metaheuristic algorithms to solve routing problems will be implemented, which would further improve the quality of created routes and optimization capabilities. At the same time, due to multiple routing solutions for the same customers and different vehicles, the use of parallelization and storage of solutions in memory can lead to improvements and acceleration of the resolution process.

References

1. Dantzig, G.B., Ramser, J.H.: The truck dispatching problem. Manage. Sci. **6**(1), 80–91 (1959). https://doi.org/10.1287/mnsc.6.1.80
2. Yang, X.S.: Nature-Inspired Metaheuristic Algorithms. Luniver Press, Frome (2010)
3. Tan, Y., Zhu, Y.: Fireworks algorithm for optimization. In: Tan, Y., Shi, Y., Tan, K.C. (eds.) ICSI 2010. LNCS, vol. 6145, pp. 355–364. Springer, Heidelberg (2010). https://doi.org/10.1007/978-3-642-13495-1_44
4. Dorigo, M., Di Caro, G.: Ant colony optimization: a new meta-heuristic. In: Proceedings of the Congress on Evolutionary Computation (CEC 1999) (Cat. No. 99TH8406), vol. 2, pp. 1470–1477 (1999). https://doi.org/10.1109/CEC.1999.782657
5. Wang, G.G., Deb, S., Coelho, L.D.S.: Elephant herding optimization. In: 3rd International Symposium on Computational and Business Intelligence (ISCBI), pp. 1–5 (2015). https://doi.org/10.1109/ISCBI.2015.8
6. Žunić, E., Delalić, S., Hodžić, K., Beširević, A., Hindija, H.: Smart warehouse management system concept with implementation. In: 14th Symposium on Neural Networks and Applications (NEUREL), pp. 1–5 (2018). https://doi.org/10.1109/NEUREL.2018.8587004
7. Zunic, E., Delalic, S., Tucakovic, Z., Hodzic, K., Besirevic, A.: Innovative modular approach based on vehicle routing problem and ant colony optimization for order splitting in real warehouses. In: 14th Conference on Computer Science and Information Systems (FedCSIS) (Communication Papers), pp. 125–129 (2019). https://doi.org/10.15439/2019f196
8. Delalic, S., Zunic, E., Alihodzic, A., Selmanovic, E.: The order batching concept implemented in real smart warehouse. In: 43rd International Convention on Information and Communication Technology, Electronics and Microelectronics (Mipro), pp. 1685–1690 (2020). https://doi.org/10.23919/mipro48935.2020.9245256
9. Ziemba, E., Oblak, I.: Critical success factors for ERP systems implementation in public administration. Interdisc. J. Inf. Knowl. Manag. **8**, 1–19 (2013). https://doi.org/10.28945/1785
10. Ziemba, E., Eisenbardt, M.: Business processes improvement by using consumers' knowledge. Problemy Zarzadzania **15**(71), 102–115 (2017). https://doi.org/10.7172/1644-9584.71.7
11. Alihodzic, A., Delalic, S., Gusic, D.: An effective integrated metaheuristic algorithm for solving engineering problems. In: 15th Conference on Computer Science and Information Systems (FedCSIS), pp. 207–214 (2020). https://doi.org/10.15439/2020F81

12. Žunić, E., Beširević, A., Delalić, S., Hodžić, K., Hasić, H.: A generic approach for order picking optimization process in different warehouse layouts. In: 41st International Convention on Information and Communication Technology, Electronics and Microelectronics (Mipro), pp. 1000–1005 (2018). https://doi.org/10.23919/MIPRO.2018.8400183
13. Osaba, E., Yang, X.S., Diaz, F., Lopez-Garcia, P., Carballedo, R.: An improved discrete bat algorithm for symmetric and asymmetric traveling salesman problems. Eng. Appl. Artif. Intell. **48**, 59–71 (2016). https://doi.org/10.1016/j.engappai.2015.10.006
14. Saji, Y., Riffi, M.E.: A novel discrete bat algorithm for solving the travelling salesman problem. Neural Comput. Appl. **27**(7), 1853–1866 (2016). https://doi.org/10.1007/s00521-015-1978-9
15. Jati, G.K., Suyanto: Evolutionary discrete firefly algorithm for travelling salesman problem. In: Bouchachia, A. (ed.) ICAIS 2011. LNCS, vol. 6943, pp. 393–403. Springer, Heidelberg (2011). https://doi.org/10.1007/978-3-642-23857-4_38
16. Taidi, Z., Benameur, L., Chentoufi, J.A.: A fireworks algorithm for solving travelling salesman problem. Int. J. Comput. Syst. Eng. **3**(3), 157–162 (2017). https://doi.org/10.1504/ijcsyse.2017.086740
17. Hossam, A., Bouzidi, A., Riffi, M.E.: Elephants herding optimization for solving the travelling salesman problem. In: Ezziyyani, M. (ed.) AI2SD 2018. AISC, vol. 912, pp. 122–130. Springer, Cham (2019). https://doi.org/10.1007/978-3-030-12065-8_12
18. Wang, K.P., Huang, L., Zhou, C.G., Pang, W.: Particle swarm optimization for traveling salesman problem. In: Proceedings of the International Conference on Machine Learning and Cybernetics (Cat. No. 03ex693), vol. 3, pp. 1583–1585 (2003). https://doi.org/10.1109/icmlc.2003.1259748
19. Baker, B.M., Ayechew, M.A.: A genetic algorithm for the vehicle routing problem. Comput. Oper. Res. **30**(5), 787–800 (2003). https://doi.org/10.1016/S0305-0548(02)00051-5
20. Chiang, W.C., Russell, R.A.: Simulated annealing metaheuristics for the vehicle routing problem with time windows. Ann. Oper. Res. **63**(1), 3–27 (1996). https://doi.org/10.1007/BF02601637
21. Gendreau, M., Hertz, A., Laporte, G.: A tabu search heuristic for the vehicle routing problem. Manage. Sci. **40**(10), 1276–1290 (1994). https://doi.org/10.1287/mnsc.40.10.1276
22. Caceres-Cruz, J., Arias, P., Guimarans, D., Riera, D., Juan, A.A.: Rich vehicle routing problem: Survey. ACM Computing Surveys (CSUR) **47**(2), 1–28 (2014). https://doi.org/10.1145/2666003
23. Osaba, E., Yang, X. S., Del Ser, J.: Is the Vehicle Routing Problem Dead? An Overview Through Bioinspired Perspective and a Prospect of Opportunities. In Nature-Inspired Computation in Navigation and Routing Problems, pp. 57–84 (2020). https://doi.org/10.1007/978-981-15-1842-3_3
24. Ai, T.J., Kachitvichyanukul, V.: A particle swarm optimization for the vehicle routing problem with simultaneous pickup and delivery. Computers & Operations Research **36**(5), 1693–1702 (2009). https://doi.org/10.1016/j.cor.2008.04.003
25. Chen, A.L., Yang, G.K., Wu, Z.M.: Hybrid discrete particle swarm optimization algorithm for capacitated vehicle routing problem. Journal of Zhejiang University-Science A **7**(4), 607–614 (2006). https://doi.org/10.1631/jzus.2006.A0607

26. Gong, Y.J., Zhang, J., Liu, O., Huang, R.Z., Chung, H.S.H., Shi, Y.H.: Optimizing the vehicle routing problem with time windows: a discrete particle swarm optimization approach. IEEE Transactions on Systems, Man, and Cybernetics, Part C (Applications and Reviews), 42(2), 254–267 (2012). https://doi.org/10.1109/TSMCC.2011.2148712
27. Belmecheri, F., Prins, C., Yalaoui, F., Amodeo, L.: Particle swarm optimization algorithm for a vehicle routing problem with heterogeneous fleet, mixed backhauls, and time windows. J. Intell. Manuf. **24**(4), 775–789 (2013). https://doi.org/10.1007/s10845-012-0627-8
28. Taha, A., Hachimi, M., Moudden, A.: Adapted bat algorithm for capacitated vehicle routing problem. International Review on Computers and Software (IRECOS) **10**(6), 610–619 (2015). https://doi.org/10.15866/irecos.v10i6.6512
29. Zhou, Y., Luo, Q., Xie, J., Zheng, H.: A hybrid bat algorithm with path relinking for the capacitated vehicle routing problem. In Metaheuristics and Optimization in Civil Engineering, pp. 255–276 (2016). https://doi.org/10.1007/978-3-319-26245-1_12
30. Osaba, E., Carballedo, R., Yang, X.S., Fister Jr, I., Lopez-Garcia, P., Del Ser, J.: On efficiently solving the vehicle routing problem with time windows using the bat algorithm with random reinsertion operators. In Nature-Inspired Algorithms and Applied Optimization, pp. 69–89 (2018). https://doi.org/10.1007/978-3-319-67669-2_4
31. Cai, Y., Qi, Y., Cai, H., Huang, H., Chen, H.: Chaotic discrete bat algorithm for capacitated vehicle routing problem. International Journal of Autonomous and Adaptive Communications Systems **12**(2), 91–108 (2019). https://doi.org/10.1504/IJAACS.2019.098589
32. Bell, J.E., McMullen, P.R.: Ant colony optimization techniques for the vehicle routing problem. Adv. Eng. Inform. **18**(1), 41–48 (2004). https://doi.org/10.1016/j.aei.2004.07.001
33. Yu, B., Yang, Z.Z., Yao, B.: An improved ant colony optimization for vehicle routing problem. Eur. J. Oper. Res. **196**(1), 171–176 (2009). https://doi.org/10.1016/j.ejor.2008.02.028
34. Szeto, W.Y., Wu, Y., Ho, S.C.: An artificial bee colony algorithm for the capacitated vehicle routing problem. Eur. J. Oper. Res. **215**(1), 126–135 (2011). https://doi.org/10.1016/j.ejor.2011.06.006
35. Yao, B., Hu, P., Zhang, M., Wang, S.: Artificial bee colony algorithm with scanning strategy for the periodic vehicle routing problem. SIMULATION **89**(6), 762–770 (2013). https://doi.org/10.1177/0037549713481503
36. Korayem, L., Khorsid, M., Kassem, S. S.: Using grey wolf algorithm to solve the capacitated vehicle routing problem. In IOP Conference Series: Materials Science and Engineering. IOP Publishing, 83(1), 1–10 (2015). https://doi.org/10.1088/1757-899X/83/1/012014
37. Yang, W., Ke, L.: An improved fireworks algorithm for the capacitated vehicle routing problem. Frontiers of Computer Science **13**(3), 552–564 (2019). https://doi.org/10.1007/s11704-017-6418-9
38. Osaba, E., Yang, X.S., Diaz, F., Onieva, E., Masegosa, A.D., Perallos, A.: A discrete firefly algorithm to solve a rich vehicle routing problem modelling a newspaper distribution system with recycling policy. Soft. Comput. **21**(18), 5295–5308 (2017). https://doi.org/10.1007/s00500-016-2114-1
39. Altabeeb, A.M., Mohsen, A.M., Ghallab, A.: An improved hybrid firefly algorithm for capacitated vehicle routing problem. Applied Soft Computing **84**, 1–9 (2019). https://doi.org/10.1016/j.asoc.2019.105728

40. Osaba, E., Carballedo, R., Yang, X.S., Diaz, F.: An evolutionary discrete firefly algorithm with novel operators for solving the vehicle routing problem with time windows. In Nature-Inspired Computation in Engineering, pp. 21–41 (2016). https://doi.org/10.1007/978-3-319-30235-5_2
41. Matthopoulos, P.P., Sofianopoulou, S.: A firefly algorithm for the heterogeneous fixed fleet vehicle routing problem. Int. J. Ind. Syst. Eng. **33**(2), 204–224 (2019). https://doi.org/10.1504/IJISE.2019.102471
42. Vidal, T., Battarra, M., Subramanian, A., Erdogan, G.: Hybrid metaheuristics for the clustered vehicle routing problem. Computers & Operations Research **58**, 87–99 (2015). https://doi.org/10.1016/j.cor.2014.10.019
43. Dondo, R., Cerdá, J.: A cluster-based optimization approach for the multi-depot heterogeneous fleet vehicle routing problem with time windows. Eur. J. Oper. Res. **176**(3), 1478–1507 (2007). https://doi.org/10.1016/j.ejor.2004.07.077
44. Expósito-Izquierdo, C., Rossi, A., Sevaux, M.: A two-level solution approach to solve the clustered capacitated vehicle routing problem. Computers & Industrial Engineering **91**, 274–289 (2016). https://doi.org/10.1016/j.cie.2015.11.022
45. Zunic, E., Donko, D., Supic, H., Delalic, S.: Cluster-based approach for successful solving real-world vehicle routing problems. In 15th Conference on Computer Science and Information Systems (FedCSIS), Vol. 21, pp. 619–626 (2020). https://doi.org/10.15439/2020F184
46. Alsabti, K., Ranka, S., Singh, V.: An efficient k-means clustering algorithm. (1997)
47. Park, H.S., Jun, C.H.: A simple and fast algorithm for K-medoids clustering. Expert Systems with Applications **36**(2), 3336–3341 (2009). https://doi.org/10.1016/j.eswa.2008.01.039
48. Johnson, S.C.: Hierarchical clustering schemes. Psychometrika **32**(3), 241–254 (1967). https://doi.org/10.1007/BF02289588
49. Senthilnath, J., Omkar, S.N., Mani, V.: Clustering using firefly algorithm: performance study. Swarm and Evolutionary Computation **1**(3), 164–171 (2011). https://doi.org/10.1016/j.swevo.2011.06.003
50. Xie, H., Zhang, L., Lim, C.P., Yu, Y., Liu, C., Liu, H., Walters, J.: Improving K-means clustering with enhanced Firefly Algorithms. Applied Soft Computing **84**, 1–22 (2019). https://doi.org/10.1016/j.asoc.2019.105763
51. Van der Merwe, D. W., Engelbrecht, A. P.: Data clustering using particle swarm optimization. In Congress on Evolutionary Computation (CEC'03), Vol. 1, pp. 215–220 (2003). https://doi.org/10.1109/CEC.2003.1299577
52. Kudova, P.: Clustering genetic algorithm. In 18th International Workshop on Database and Expert Systems Applications (DEXA), pp. 138–142 (2007). https://doi.org/10.1109/DEXA.2007.65
53. Selim, S.Z., Alsultan, K.: A simulated annealing algorithm for the clustering problem. Pattern Recogn. **24**(10), 1003–1008 (1991). https://doi.org/10.1016/0031-3203(91)90097-O
54. Žunić, E., Delalić, S., Hodžić, K., Tucaković, Z.: Innovative GPS Data Anomaly Detection Algorithm inspired by QRS Complex Detection Algorithms in ECG Signals. In 18th International Conference on Smart Technologies (EUROCON), pp. 1–6 (2019). https://doi.org/10.1109/EUROCON.2019.8861619
55. Žunić, E., Hindija, H., Beširević, A., Hodžić, K., Delalić, S.: Improving Performance of Vehicle Routing Algorithms using GPS Data. In 14th Symposium on Neural Networks and Applications (NEUREL), pp. 1–4 (2018). https://doi.org/10.1109/NEUREL.2018.8586982

56. Žunic, E., Kuric, A., Delalic, S.: Improving unloading time prediction for Vehicle Routing Problem based on GPS data. In 15th Conference on Computer Science and Information Systems (FedCSIS) (Position Papers), Vol. 22, pp. 45–49 (2020). https://doi.org/10.15439/2020f123
57. Žunić, E., Delalić, S., Donko, D.: Adaptive multi-phase approach for solving the realistic vehicle routing problems in logistics with innovative comparison method for evaluation based on real GPS data. Transportation Letters **1–14**, (2020). https://doi.org/10.1080/19427867.2020.1824311
58. Yang, X.S., He, X.: Firefly algorithm: recent advances and applications. International Journal of Swarm Intelligence **1**(1), 36–50 (2013). https://doi.org/10.1504/ijsi.2013.055801
59. Žunić, E., Delalić, S., Donko, Dž., Šupić, H.: Two-phase Approach for Solving the Rich Vehicle Routing Problem Based on Firefly Algorithm Clustering. In 6th International Congress on Information and Communication Technology (ICICT). (2021). Paper accepted

Technological Infrastructure for Business Excellence

Street Addressing System as an Essential Component in Sustainable Development and Realization of Smart City Conception

Dmitriy Gakh(✉)

Institute of Control Systems of ANAS, Bakhtiyar Vahabzadeh Street 68, AZ1141 Baku, Azerbaijan
dgakh@sinam.net

Abstract. Although Street Addressing (SA) seems to be a well-known conception, there is misunderstanding of it and its contribution to the Urban Economics, Smart Cities (SC) initiatives, and Sustainable Development (SD). The role of SA in planning, building, and operation of SC remains unclear. Considering SA as a socio-technical system and underlying part of ICT services allowed formulating peculiarities of SA system towards the implementation of the SC conception [1]. This paper expands the SA for SC research and includes considering SA contribution to SD. Discovery of indirect influence of SA peculiarities to SD is the main and the most important result of this article. This discovery can be considered as an example of SA Quality Assurance for actual SA program and in some extent fills the research gap in SA methodology. The goal of this article is an analysis of how SA affects SC initiatives and SD. Uniqueness of this research is concluded in analysis of SA as a urban infrastructure component through the prism of Software Quality that enables building quality SC services. This approach can be considered as a contribution to science because it shows a possibility to synthesize the ICT and urban management/civil engineering. It also highlights a role of the government in developing and implementing the SA programs as well as issues of integration of SA with urban infrastructure identification and inventory systems. The only limitation could be the use of the available SC literature that is not sufficient to confidently draw long-term conclusions.

Keywords: Street addressing · Smart city · Sustainable development

1 Introduction

A Street Addressing (SA) system is an essential component of a modern city. Although SA is a well-known conception, a lot of problems relate to misunderstanding of what it really is. Such problems are rooted in the fact that the complexity of modern cities is growing while conception of SA is not being changed. Development of Information and Communication Technologies (ICTs) and increased urbanization are grounded in conception of so-called Smart City (SC). There is no established definition of what a SC is, rather SC can be considered as direction of city development. SC conception is young

E. Ziemba and W. Chmielarz (Eds.): ISM 2020/FedCSIS-IST 2020, LNBIP 413, pp. 127–145, 2021.
https://doi.org/10.1007/978-3-030-71846-6_7

– publications about smart cities are dated 2012. There is also lack of proper research and publications about SA [1].

The gap between urbanization and technological development and lack of modern SA research cause new problems relating to implementation of new technologies without proper research. Such problems arise because a small change in technology is available at the business level, which is interested in the speedy implementation of the solution without an in-depth laboratory analysis of the consequences [1–3]. As a result, cases related to such implementation of the solutions are not reflected by literature and do not contribute science. Discussion of this problem leads to philosophy and makes questions for another research. Study of the situation with the SA can disclose interesting facts and contribute researchers with new findings.

This paper considers contribution of SA systems to sustainable urban development. Although many peculiarities of SA systems (PSAS) do not show a direct influence of SA to SC, there may be an indirect affect [1]. This article aims to discover such influence and describe it as an example that can be used in actual SA programs. At the same time this research contribute science by an example of synthesis of ICTs and civil engineering and showing how urban management/civil engineering can be driven by the ICT quality assurance approaches. This research is unique because there was no similar approach in studying of SA systems and Sustainable Development (SD) through prism of Software Quality.

Movement of cities to SC conception is also considered from point of view of sustainable urban development process. Research presented in this paper is based on report, presented in FedCSIS 2020 Conference [1] and expands this report by additional findings. As it was reported in [1], SA systems are considered as a part of Information Systems providing related service. PSAS were discussed in context of movement towards SC conception. This paper in its turn is aimed to consider PSAS in context of SD.

This research is based on literature analysis of three main types of papers: SA related, SC related, and SD related papers. Independence of the topics discussed in these papers reflects the scientific gap. The lack of literature describing how SA system affects SC and DS is explicit evidence of this gap. The problem discussed in this article is concluded in absence of researching into the relations between SD and SA. This absence does not allow to understand the SA systems' influence on SD and how important it is. SA related and SC related papers were used for development of PSAS [1]. Finally, this paper refers PSAS and SD related papers and synthesizes findings of all three types of papers. The use of the available SC literature is not sufficient to confidently draw long-term conclusions because rapid development of ICTs and SC. At the same time indirect influence of PSAS to SD was discovered and discussed that could be the main and the most important result of this article.

The chapter "Main Literature on Street Addressing, Smart Cities, and Sustainable Development" describes literature analysis in more details. The chapter "Street Addressing for Smart Cities" contains important exclusions from the research of PSAS. Research questions are presented in eponymous chapter. The chapter "Research Methodology" describes the research itself. The chapter "Research Findings" includes important outcomes of the literature analysis. The chapter "Discussion of Findings" contains discussion of important outcomes through the prism of ecological, economic, socio-cultural, and political aspects of SD. Findings related to mobility, city administration, and knowledge management are also discussed in this chapter. In addition, it is an attempt to define variables. The chapter "Conclusion" summarizes this article.

2 Theoretical Background and Research Questions

2.1 Main Literature on Street Addressing, Smart Cities, and Sustainable Development

Studies of the World Bank [4] can be considered as the main source for study of SA. It contains methodology of SA programs design and implementation, description of many cases of SA related issues for cities in developing countries. The most important findings presented in this source are advantages bringing a SA system for cities which has no SA system in place. At the same time this book contains different types of SA systems reducing the SA programs cost, and criterion of SA system type selection for specific cases. The mentioned book can be safely considered as an ABC in SA.

Another source of information includes written sources about SC. Although there is lack of such literature in scientific sources, existing articles contain sufficient information to understand that SC conception is and form solid base for this research.

The third source of information that was used in this analysis is literature about Software Quality (SQ). Considering SA systems as services that can be implemented by means of ICTs, solutions principles of SQ can be applied and expanded to the SA systems. This approach is a key finding of research presented in this paper and in [1]. Expansion of SQ principles to SA solutions is not simple and requires synthesis of solid knowledge of ICTs and urban infrastructure/urban management/civil engineering.

The fourth source of information is literature about SD. This source includes, for example, [5], where ecological, economic, socio-cultural, and political components are proposed as those of sustainability. Sustainable Information Society (SIS) is assumed as one of the constituent parts of SD. ICTs outlay, information culture, ICTs management, and ICTs quality are considered as components influencing ICTs adoption in [6]. PSAS will be discussed below through the prism of these components.

Social Adaptive Case Management (SACM) is discussed in [7]. SACM approaches are considered as ones that can potentially be used to develop SA services. This approach to Knowledge Management is interesting in application to SA system establishment and operation from point of view of Informational Society development.

The turning point for civilization, which was initiated in 1960's, is referred to society's communication via computer, post-industrial society, the third wave, post-capitalism society and, finally, the information society. This is related to the increasing role of information and ICTs in creating modern society. The creation and development of modern society do not solemnly rely on material and financial resources but are also based on an intangible asset, which is information [8].

Those which have the greatest influence on the development of the SIS include [8]:

- The growing importance of knowledge and knowledge management;
- The rapid development of ICTs and accelerating digitization;
- The increasing globalization;

- The networking and design approach development; and
- The emergence of generations Y/Z/Alpha, namely a new generation of students, employees and consumers.

2.2 Street Addressing for Smart Cities

This chapter includes the main theses from [1] that describe the SA for SC issues and are essential to set questions for this research. Although SA system is an essential component of the city infrastructure, all problems concerning the SA are not currently reflected by proper research or sufficiently described. The lack of research and resources is observed especially in SA issues relating to implementation of SC conception.

PSAS that can be used for planning, implementing, and maintaining the SA system focused on implementing the SC conception were determined and defined [1]. Although the meaning of Smart City is not settled yet, there is an agreement on the significant role of ICTs in smart urban development [9]. SC citizens are considered as "prosumers" of geo-tagged data and content affecting cities' everyday norms and interactions [10]. The detail study shows that SC can be considered as a socio-technical system [11]. It means that the citizens form some kind of information society.

Although SA seems to be a well-known and well-studied conception that does not require research, this is not true. First of all, there are many cities worldwide implementing SA program [12]. Experience gained in these programs forms solid base for research. Secondly, implemented SA programs are targeted mainly to solve existing problems, but do not focus on the future technologies. Indeed, the absence of the SA system has rather greater negative effect to the urban economy than positive impact that could bring orientation towards the SC. Cost of the SA system and planning the faster Return of Investments (ROI) are additional factors leading to avoid such orientation.

Implementation of ICTs in modern cities creates new environment to process city-related data. Optimization of physical city infrastructure, growth, concentration, and decentralization of this infrastructure introduce new requirements to identify location of the infrastructure units (urban fixtures). An example of problems relating to the increase of density within the urban areas are street vendors issues. In poor areas the street vendors problems are complex and include lack of territorial management, poor taxation, and legal violations [13]. Proper ICTs application and SA system could mitigate the problems by increasing manageability of territories where street vendors can be deployed.

New construction technologies are changing shape of cities today. Smart infrastructure and architecture are dominating the shift in the construction industry. In the Smart Cities Connect Conference and Expo in 2017, Skansa USA Chief Sustainability Officer Elizabeth Heider highlighted resilient design and sustainable construction as key aspects of how the construction industry can adapt to the needs of the Smart City initiatives in Santa Monica. A recent survey cited at ReadWrite notes that at least 60% of builders are now aware of Internet of Things (IoT) technologies and another 43% say that IoT will shape the vision of future construction. By integrating IoT into the building process, including data capture and analytics, the new infrastructures built are future proof [14]. These facts prove the increase of density and granularity of urban infrastructure and leading role of ICTs in construction.

The SA system should be able to provide services directly to people in convenient manner without special means (this requirement was considered above in this paper). There are cases when existing SA systems are obsolete and no longer meet such requirements. For example, such problems occurred in Baku, Azerbaijan where old SA system was being upgraded by the SA program [12].

Several attempts were made to develop a universal SA system or SA system for SC. These attempts were not successful for several reasons. First of all, SA and Postal Addresses are different conceptions. Authors describe their systems as SA although their systems were based on Postal Services requirements. Another problem is an attempt to build complex code for the address to reduce its readability.

Commerce-Science Problem can be understood by examples of Max Tegmark's letter and "what3words" service (https://www.what3words.com). The problem is rooted in the fact that a small change in technology is available at the business level, interested in the speedy implementation of the solution without proper research [2, 3].

2.3 Research Questions

Issues of SA for SC were discussed in [1] that constitutes the foundation to development new SA methodologies. SC and SA consequently were considered as a socio-technical system. This paper considers SA for SC from point of view of SD. As a result, this research addresses the following research questions:

RQ1: Do the SA systems contribute to SD and how?
RQ2: Which peculiarities of the SA systems contribute to SD and how?
RQ3: What role does the SA system play in SC and SD?

3 Research Methodology

Research of SA for SC is extremely complex as it requires to study many implicit factors and interrelations. Scope and extent of this survey is city-wide. The research is trying to make sense of the myriad of human, political, social, cultural and contextual elements involved.

This research implies in descriptive design. Synthesis of SQ and urban infrastructure/urban management/civil engineering methodologies and subsequent description of PSAS were described in [1]. This paper considers evaluation of influence of PSAS to SD. Both holistic approach towards the complex interplay of many indicators and responsive evaluation – which entails a series of investigative steps to evaluate how responsive a situation is to all those involved. A common purpose of the evaluation is to examine the projects performance from the point of view of awareness levels, costs and benefits, cost-effectiveness, attainment of objectives and quality assurance. The results are used to prescribe changes to improve and develop the situation. The research is guided by [15].

This research is based on literature analyses and in 12-years practical implementation of projects in Software Development, Quality Assurance, Geographic Information Systems (GIS), development of navigation solutions for urban areas of Azerbaijan and

Georgia (www.GoMap.Az project), and development of methodology for SA program for Kabul, Afghanistan.

Several components of SD are selected for this research. PSAS described in [1] were analyzed thought the prism of these components. Although SD is a quite complex subject, we can assume that this research covers the core issues of how SA contributes to SD. So, this research can be considered as a starting point for study and planning of SA programs towards SD.

According to abstraction levels [15] this paper, as well as previous papers describing research of SA in application to SC [1, 3] can be as follows:

- Theory. Theoretical aspect of this research is synthesis of two main subjects – SA and SQ. SA in this context is considered as a service that can be implemented by means of ICTs. This approach requires, first of all, theoretical studies of SA and SQ. Thus, this research describes the theory;
- Concepts. Concepts, discussed in this research are PSAS, mainly represented by so-called "-ilities" like "-ilities" in SQ;
- Indicators. Depth of this research allows considering some indicators, but not all of them. Indicators are described tightly with concepts and variables;
- Variables. There is a trial to determine some variables is this research. Recommendations of how to measure corresponding components are also considered;
- Values. This research is a theory and it does not consider values.

Theoretical statements of this research should be proven in practice. The best way to determine values is development of SA programs for specific cases. This research is positioned as a foundation theory and can be considered as a basis for future research as well as just recommendations. Problems described in this paper are of interest of scientists who perform research in modern urban development.

4 Research Findings

Research described in [1] and this research include analyses and build the theory covering the following subjects:

- Urban Infrastructure/Urban Management/Civil Engineering (literature and experience in actual SA programs);
- Smart City Conception (literature);
- Software Development/Software Quality (literature and experience in actual Software Development);
- Sustainable Development (literature).

Development of ICTs should be considered as a main driver for subjects discussed. This development leads appearance of Smart City Conception that impacts the urban infrastructure, urban management, and civil engineering. Being a part of ICTs, Software Development and Software Quality contribute to this impact. SA is an essential component of realization of SC conception [1], ICTs is a driver of development of both SC conception and Information Society. SC is a socio-technical system [1]. Thus, one can say that in some extent SA impacts Information Society.

4.1 Sustainable Information Society

Delgado states the following specific facts about modern Web [16]:

- People evolved from mere browsing and information retrieval to first class players, either by actively participating in business workflows (Business Web) or engaging in leisure and social activities (Social Web);
- The service paradigm became widespread, in which each resource (electronic or human) can both consume and provide services, interacting in a global Service Web;
- Platforms are now virtualized, dynamic, elastic and adaptable, cloud-style, with mobility and resource migration playing a role with ever increasing relevance;
- The distributed system landscape has expanded dramatically, both in scale (with a massive number of interconnected nodes) and granularity (nodes are now very small embedded computers), paving the way for the Internet of Things (IoT).

These facts are the evidence that people and IoT devices play role of prosumers (providers and consumers) and that Informational Systems become global, distributed and granulated. For SC conception we can say that citizens form information society.

The sustainable information society (SIS) is a new phase of information society development in which ICTs are becoming key enablers of sustainability. Overall, the SIS is a multidimensional concept encompassing environmental, economic, cultural, social, and political aspects, all of which could be strongly influenced by adopting ICTs by society stakeholders, mainly enterprises, households, and public administration. Some researchers have identified ICTs as one of the most important tools in building sustainable business practices and supporting the success of businesses. Moreover, the ICTs adoption in enterprises can yield benefits in environmental preservation by increasing energy efficiency and equipment utilization as well as it can influence social development by making information available to all society stakeholders. All these possibilities make ICTs enablers of sustainability in several respects, i.e. environmental protection (ecological sustainability), economic growth (economic sustainability), socio-cultural development (socio-cultural sustainability), and governance (political sustainability) [5].

According to Schauer [17], SD has four dimensions which are ecological, social, economic and cultural sustainability. In a further study, Ziemba proposed an expanded sustainability which included four sustainability components, i.e. ecological, economic, socio-cultural, and political [5].

Regarding businesses, the sustainability components are [5]:

- Ecological sustainability is the ability of enterprises to maintain rates of renewable resource harvest, pollution creation, and non-renewable resource depletion by means of conservation and proper use of air, water, and land resources;
- Economic sustainability of enterprises means that enterprises can gain competitive edge, increase their market share, and boost shareholder value by adopting sustainable practices and models. Among the core drivers of a business case for sustainability are: cost and cost reduction, sales and profit margin, reputation and brand value, innovative capabilities;

- Socio-cultural sustainability is based on the socio-cultural aspects that need to be sustained, e.g. trust, common meaning, diversity as well as capacity for learning and capacity for self-organization. It is seen as dependent on social networks, making community contributions, creating a sense of place and offering community stability and security; and
- Political sustainability must rest on the basic values of democracy and effective appropriation of all rights. It is related to the engagement of enterprises in creating democratic society.

Based on a stream of research, Ziemba advanced a model of SIS in which the ICTs adoption construct is composed of four sub-constructs: ICTs outlay, information culture, ICTs management, and ICTs quality. The sub-construct of ICTs outlay includes the government units' financial capabilities and expenditure on the ICTs adoption. The information culture subconstruct embraces digital and socio-cultural competences of government units' employees and managers, constant improvement of these competences, employees' personal mastery and creativity, and incentive systems encouraging employees to adopt ICTs. The ICTs management sub-construct comprises the alignment between information society strategy and ICTs adoption, top management support for ICTs projects, as well as the adoption of newest management concepts and standard ICTs solutions developed at the national level. It also includes the implementation of legal regulations associated with the ICTs adoption, regulations on ICTs and information security and protection [6].

The ICTs quality sub-construct consists of the quality, interoperability, and security of back- and front-office information systems, quality of hardware, maturity of e-public services, and the adoption of electronic document management system, electronic delivery box, as well as ERP and BI systems. The construct asserted that the four sub-constructs were interrelated and critical to the design of the ICTs adoption in government units in the context of the SIS. The definition of SD employed throughout this paper relates to a development in which the needs of present generations are met without compromising the chances of future generations to meet their own needs [6].

PSAS, presented in [1] are elaborated on base of synthesis of SQ and urban infrastructure/urban management issues. This fact shows that PSAS can be understandable for both civil engineers and software developers. So, above mentioned sub-constructs i.e. ICTs outlay, information culture, ICTs management, and ICTs quality can be easily considered along with PSAS. So, PSAS can be assessed in terms of SIS development. Cost, Aesthetics (additional peculiarity that will be discussed below), Manageability, Efficiency, Usability are examples of peculiarities that seems to be close in relation with four sub-constructs mentioned.

Open Source Software (OSS) has been identified as a strategy for implementing long-term sustainable software systems [18]. For any OSS project, the sustainability of its communities is fundamental to its long-term success [19]. Taking into account that users of SA services become prosumers, they form community of SA or related services. Thus, one can claim that SA service based on open data, open structures, and open algorithms is fundamental to its long-term success and contributes to the political dimension of sustainable dimension.

4.2 Smart City Initiatives and Sustainable Development

Federico Caprotti states that the City Council has, itself, highlighted the barriers it sees in rolling out a smart city/digital city strategy in Birmingham. Among other the identified barriers include fragmented and incomplete GIS information about city utilities and the slow pace and bureaucratic nature of civic planning processes [20]. GIS problems can be addressed by Structural SA quality and Addressability, Correctness, Integrity (the PSAS).

Three of the initiatives forming part of the Future City Glasgow Program are briefly described below: the Glasgow Operations Centre; Intelligent Street Lighting; and a series of Community Mapping events. The first one represents the most substantial tangible legacy of the Program; the other two exemplify the varied smaller-scale or short-lived activities surrounding it. This wide range of additional activities implemented includes [20]:

- My Glasgow (an app allowing citizens to report problems);
- Linked Mapping (an online map including listings of various amenities and services);
- Renewables Mapping (a map of renewable energy opportunities within the city);
- Dashboards (allowing users to choose which city datasets will be visualised);
- Active Travel (a demonstrator smart phone app designed for pedestrians and cyclists);
- Hacking the Future (a series of four public 'hackathons' themed around public safety, energy, health, and transport);
- Energy Efficiency Demonstrator (creating a detailed data-based portrait of energy consumption across the city, to investigate the potential to reduce carbon emissions, lower fuel bills, and address fuel poverty);
- Social Transport (smart phone software allowing the city's social transport fleet to provide an improved, more flexible service);
- Engaging the City (a touring exhibition stands to raise awareness of the Future City initiative);
- Future Makers (a series of events teaching children to code);
- Citizen Engagement (a research study investigating citizens' views, with a focus on waste and road repairs).

This list shows that many SA-relating activities can be implemented as ICT services. At the same time these activities are evidence of information society development. Considering quality of the SA systems through peculiarities discussed in [1] allows providing quality assurance required for corresponding ICT solution because these peculiarities are based on synthesis of SQ and urban infrastructure/urban management/civil engineering issues (as it was mentioned above).

The new functionalities being tested in Glasgow included: sensors providing real-time data on sound levels, air quality, and pedestrian footfall; and 'Dynamic' lights able to detect motion and raise lighting levels accordingly. The data are fed automatically into the city's open data hub, launched in February 2015 [20]. Addressability is the PSAS that can directly address initiative-related issues.

The 'Domains' and 'Subdomains' of the Sheffield Smart City Strategy framework [20]:

- Smart People: Welfare and social inclusion, Digital inclusion, Education, Human capital management;
- Smart Resources: Renewable energy, Smart grids, Waste management, Water management, Food & agriculture;
- Smart Mobility: Public transport, Private transport, City logistics, Road network, Public lighting;
- Smart Buildings: Facilities management, Building services (Mechanical & Electric), Housing quality, Assisted living, Cultural heritage management;
- Smart Living: Healthcare & wellbeing, Entertainment & Sport, Culture, Retail, Hospitality & Tourism, Pollution control, Public spaces management, Security (blue light);
- Smart Economy: Innovation & Entrepreneurship, Digital skills, eBusiness;
- Smart Governance: e-Government, e-Democracy, Procurement, Transparency, Communications.

Taking into account that SA services are operated by geo-location, it can be said that SA explicitly impacts the components of the Sheffield Smart City Strategy framework such as Smart grids, Waste management, Water management, Public transport, Private transport, City logistics, Road network, Public lighting, Facilities management, Building services, Housing quality, Assisted living, Cultural heritage management, Hospitality & Tourism, Public spaces management, Security, Innovation & Entrepreneurship (related to SA services), e-Business, e-Government.

At the level of the everyday life of urban communities a key issue is how to revitalize cities in order to maintain their ability to provide welfare to their citizens in a sustainable manner. The role of citizen, user, and stakeholder participation in smart environments and platforms in major Finnish cities contributes to urban economic renewal by enhancing productive smartness. Such participation has an inherent instrumental dimension but its rationale goes beyond narrow-minded instrumentalism.

Three Finnish cities, Helsinki, Tampere, and Oulu, have started to utilize participatory platforms to foster innovation, which is in line with local strategies and backed up by intercity collaboration (Six City Strategy), the national Innovative Cities (INKA) Program and various programs of the European Union. The three cases above, Helsinki, Tampere, and Oulu show that these cities provide a wide range of activities from user involvement in product development to citizens' rights to bring their concerns to open innovation platforms.

The forms and functions of citizen participation matter, for participatory processes create various forms of citizen involvement, ranging between creativity, passivity and entrenchment, which implies that participation eventually creates different categories of citizen. When participatory innovation platforms become the norm in local development, they gradually reshape the entire city. In the welfare society context democratic culture and institutionalized solidarism provides support to such a transformation, not only in the form of broad civic participation but also at the level of material relations, for asymmetries of micro-level innovation processes and their outcomes are counterweighted by macro structures of solidarity. Such conditions are conducive to the emergence of a

Smart City that is co-created by its citizens or, as we may call it, the City-as-a-Platform [21].

5 Discussion of Findings

As it was mentioned above, there are four dimensions of SD, i.e. ecological, economic, socio-cultural, and political [5]. It was showed that SA can be evaluated from functional, structural, and process qualities [1]. Taking into account findings and scope of this research, the best way to answer the research questions is to discuss how each PSAS contribute to each of the four dimensions of SD. This analysis can be implemented through the prism of functional, structural, and process qualities.

5.1 Variables and Values

Determination of possible Concepts, Indicators, and Variables according to levels of abstraction [15] is the sound approach to prove the theory. All considered "-ilities" are tightly interdependent. For example, if the SA system is incorrect, it immediately becomes useless. Improving the readability can reduce addressability. Thus, trade-offs between the features should be achieved for each specific case. Specific cultural features of the area of the SA program should also be taken into account [1].

There are variables that correspond to explicit impact to the specific dimension of SD and those that correspond to implicit impact. They can be named as "Explicit variables" and "Implicit variables". Navigatability, for example, directly impacts the environment because of gas emission generated by the means of transport (although this impact is not quite direct). Maintainability can be considered as a peculiarity, that impacts the environment indirectly. Such indirect influence can be found in the methods and materials used in maintenance operations. If the methods imply installation of plaques on trees (in developing countries, it is sometimes observed that informal signs are nailed to the living trees) or hazardous paint it will have a negative impact to the environment.

5.2 Ecological Dimension

The influence of SA on the ecological dimension is not obvious. The biggest influence of SA on the ecological dimension is an impact of transportation. Good navigatability leads to reduction of transportation trip time and consequently gas emission. Other peculiarities also influence these savings because they increase SA quality as the total. For example, weak Readability, Correctness, Reliability, and Integrity can lead to trip to the wrong location that requires additional movement and increase in time and gas emission. Good Functionality and Flexibility can provide the best path that will also reduce the environmental impact. These peculiarities also impact operation of emergency services. Operation of firefighting services and ambulance in some extent can impact environment in case of accident with ignition of hazardous materials or elimination of a dangerous epidemic.

Safety can impact environment by using safe materials for street plaques and methods of its installation. For example, the strong light reflectivity of signs in sunny days might harm some species of animals, birds and insects. It might also blind truck drivers at night time, cause accidents and impact environment if the cargo contains some dangerous goods (DG). In windy conditions, loose plaques can be blown off and pose a risk of injury to people, animals or damage to property. Weak Safety of SA inline services can lead to hacker attacks and failures that will cause implicit environmental influence through the impact of the transportation trips. Another safety threat can an ability of unauthorized change of plaque.

Addressability can have environmental impact by allowing to address water sources, solid waste collection points, toilets, and similar facilities. If these objects are addressed, people will find them easily. SA system can also include installation of plaques-pointers within the sites. Addressing of such facilities was introduced by the SA programs implemented within the World Bank projects [4]. Maintainability can impact the environment if used materials are hazardous or maintenance implies hazardous methods.

As [22] shows, data format can affect time of processing and power consumption. It means that structure of address can directly influence the complexity of corresponding data structure and also affect time of processing and power consumption by computers and smartphones. Power consumption in its time affects battery lifetime and consequently can affect CO_2 emission. Although it is a quite low value for one device, it can be significant in calculation for all users within the city.

Proposed explicit variables could be as follows:

- Number of types of addressed Environment Relating Objects (ERO/water sources, solid waste collection points, etc.);
- Number of addressed ERO;
- Percentage of ERO within all addressed objects;
- Time spent to find necessary address/route with additional tools (Smartfone, Portable Navigation Device, etc.);
- Time spent to find necessary address/route without additional tools;
- Degree of danger of used materials;
- Presence or absence of Environmental Impact Statement (EIS) in the SA program.

Proposed implicit variables could be:

- Number of SA service failures;
- Number of accidents where SA plays role.

5.3 Economic Dimension

Undoubtedly, cost and benefits of the SA system are the main Economic Dimension relating indices. Benefits of the SA system are significantly greater than its cost. Creating or upgrading the SA system requires high capital cost that can exceed the city budget in developing countries. But benefits of the SA system can be gained continuously within next years through the development of new services/businesses. It makes SA programs interested for international investment organizations. At the same time maintenance cost should be minimized to improve ROI. SA program proposed for Kabul, Afghanistan supposes creating of two SA systems – one for the City Center, including Central Business District (CBD) and another for slums [23]. Cost for the City Center was calculated according to the best reasonable SA quality. Cost for the slums in its turn was calculated as minimum to provide mainly Navigatability and Addressability in order to support emergency services, sanitation, and tax/fee collection.

Availability of a proper SA system is essential for economic development of urban area. Almost all customer-oriented billing systems and marketing studies rely to street addresses. Kabul Ministry of Informational Technologies initiates the SA program in order to improve cellular networks services [23]. Certainly, SA services will provide the city with many other benefits. Each SA program aimed to cities without addresses supposes gaining benefit from improving taxation and collection of water/electricity service fees [4].

Proposed explicit variables could be as the follows:

- Time spent to find necessary address/route with additional tools (Smartfone, Portable Navigation Device, etc.);
- Time spent to find necessary address/route without additional tools and help;
- SA program capital and operational cost;
- Availability of SA services for open related businesses (delivery, location aware software, etc.);
- Number of organizations using SA services.

Proposed implicit variables could be as the follows:

- Number of SA service failures;
- Number of accidents where SA plays role.

5.4 Socio-Cultural Dimension

The study shows that SC can be considered as a socio-technical system. In this way SA for SC should be considered from two perspectives: to serve people and to serve technologies [1]. The PSAS, proposed in [1] have been studied from engineering point of view. Considering them in socio-cultural dimension of SD reveals the need to introduce new peculiarity "Aesthetics". Aesthetics deals with beauty and artistic taste. Thus, it relates only to human, but not to technical needs.

Beauty and artistic taste could be considered within Readability, but in this case the influence of engineering approaches can lead to design that does not fit the locality style. At the same time Aesthetics can be considered as a part of Readability. But the best way is to consider Aesthetics as a separate peculiarity.

Readability, first of all, influences persons with disabilities, especially visually impaired ones. In multicultural cities Readability can influence different population groups. Kabul SA program supposes to use labels in 3 languages: Dari, Pashto, and English [23]. Navigatability can influence children and persons with mental disorder. Addressability influences connectivity and also other PSAS.

Proposed explicit variables could be as follows:

- Social rating of the SA system in general;
- Social rating of aesthetics of SA related signs;
- Social rating of aesthetics of names of streets and other places;
- The degree of its matching the style of the locality (expert evaluation);

- Time spent to find necessary address/route with additional tools (Smartfone, Portable Navigation Device, etc.);
- Time spent to find necessary address/route without additional tools;
- Number of organizations using SA services.

Proposed implicit variables could be as follows:

- Social rating of operation of city services, including emergency services.

5.5 Political Dimension

The role of the SA system in political dimension of SD is mainly concluded in improving urban management. The influence of the SA system to citizen's political activity can be considered through operation of city services and utilities. In any case this influence is implicit and varies for each specific case. For example, in Kabul, Afghanistan development of the SA program is under control of the President Administration and considered as one component of development of the city to be one of world-known capital.

Being the basis for many location relating services, the SA system plays a key role in development of relevant ICT services and Informational Society. In this context e-Government services based on SA systems is a primary technology that can impact the political dimension of SD.

Research of the impact of enhancing a performance measurement system on the Israeli police considers socio-economic ranking factor: The common assumption in literature is that socio-economic ranking may influence on the dependent variable in some researches as in politics and government and entrepreneurship [24].

Proposed explicit variables could be as follows:

- Number of e-Government services that use SA and intensity of their usage.

Proposed implicit variables could be as follows:

- Number of tourists visiting the city and countries of their origin;
- Social rating of SA relating services including city management.

5.6 Mobility

Mobility is a key component of SC conception and modern Web [16]. E-Government and M-Government are the essential technologies related to SC conception and modern Web. It is important to note that currently the so-called mobility is one of the fastest developing phenomena. Mobile marketing is currently an essential part of electronic marketing [25]. Mobility is not absolute: it is relative to the ground. Because SA systems are bound to real estate and other ground objects, mobility is relative to the SA grid. Thus, in some cases the SA system can be considered as an "implicit" essential part of mobility. Location Based Services based on SA are examples of such services.

Services, relating to real estate and urban management use the SA grid as a reference to actual facilities and places. So, E-Government and M-Government depend on SA systems in some extent. But SA was not considered in many researches relating to E-Government and M-Government. Quality of E-Government and M-Government was evaluated by citizens' perspective (for example research of adoption of M-Government in Saudi Arabia [26, 27]).

Since quality of SA systems contributes to the quality of dependent services, it should be also considered in studies of these services. SA programs in their turn should consider the future development of dependent services. Development of SA programs with consideration of proposed peculiarities [1] based on both urban infrastructure and ITC is the best way to integrate SA systems with other services based on them. Although PSAS contribute to the quality of dependent services, Flexibility determines in what extent SA systems can make this contribution. These statements are equal not only to E-Government and M-Government, but also to many other E- and M- services, such as (E-Commerce, M-Commerce, E-Democracy, M- Democracy, etc.). M-Commerce services can consider location of potential customer and propose him/her location specific information based on GPS coordinates as well as street address. Study, presented in [25] shows that according to the survey, the second significant benefit of M-Marketing was geolocation and mobile navigation (31%). Mobility was selected as one of core digital technologies affecting the retail industry in South Africa [28].

5.7 Government Administration

The SA program can be developed and implemented only within the governmental control (city administration or its equivalent). It should not be initiated beyond the governmental control while non-government companies can develop and implement SA programs only according to their needs. It could be done in explicit or implicit manner. For example, water, power and gas supply utilities in Baku, Azerbaijan have developed and implemented their own systems of consumers identification. Such phenomena take place when existing SA system does not meet the requirements (as it was practiced by the old SA system in Baku). High Addressability and Correctness of SA system will lead to building a unified system of consumers identification and avoiding additional effort to synchronize different systems. Such SA systems seem to be a better solution for implementation of SC conception.

Development of SA that can serve as a basis of identification and inventory system for different organizations require a high level of their interoperation. High Interoperability and Integrity are primary peculiarities providing the proper interoperation between different organizations. The issues of integration between Information Systems within different organizations are considered in [29]. Success factors of such projects implemented in Government Administration Units are Organizational, Processes, Environmental, Technical, and Individual [29].

5.8 Knowledge Management

Development of the SA program methodologies, implementation of actual projects, and performing research in this area will result in producing new valuable knowledge. If SA issues used to be an activity of civil engineers, today these issues involve ICT sphere as well. Interdisciplinary penetration of different technologies requires new approaches to handle corresponding problems. Methods of Knowledge Management can be useful to handle development problems. Besides, the issues related to establishment of SA operation of corresponding services is a part of city life. It means that a citizen can produce data to be consumed by SA services for maintenance or feedback. Thus, SA will also involve the social sphere.

What has long been demanded of information systems is that they move away from the prevalent control flow perspective, commonly adopted in the business process management (BPM) area, toward the data perspective [30]. Attempts to meet these expectations have led to the emergence of a new class of information systems built around the approach known as adaptive case management (ACM). Social ACM cannot only improve the efficiency of the processing methods being used, but also help expose and propagate new information swiftly, from emergent trends to best practices [7].

6 Conclusions

Development of ICTs is the main driver for development of Smart City conception and Information Society. SD can be considered through ecological, economic, socio-cultural, and political components. The SA system is an urban geo-location system that underlays many city services. The SA system for SC should be considered as a socio-technical system to be realized both as static system of street signs and ICT services. Quality of this system and services directly impacts the operation of dependent services. Urban management, logistic services, postal services, navigation, and so on are dependent on SA services. So, the SA system is an essential component for SD and realization of SC conception.

Contribution of SA System in SD depends on a specific case. For cities without SA system development of many urban services is significantly difficult or impossible. In cities where SA was established without orientation to SC conception some services may not be effective. But for some cases established SA ceases to meet the requirements and becomes ineffective. Development of SA program with analyses of quality by peculiarities proposed in [1] is a way to achieve the best quality and build a solid foundation to develop dependent services. This analysis allows determining how SA system contributes to SD.

An attempt to discover variables for the PSAS is an example of showing how the peculiarities contribute to SD. The peculiarities are highly interdependent [1]. It might be seen that this interdependency does not allow assessing how specific peculiarity influence SD. But the value of the PSAS is an ability to analyze SA quality towards SD.

There is a significant misunderstanding of what SA is and how it impacts the city development [1]. Studying of the PSAS for each specific case allows understanding the value of SA for SD.

This paper describes theory that should be proven in practice. Practical studies are available within the actual SA programs. Knowledge Management can be used during development of methodologies for SA programs, during SA programs implementation stage, and during exploitation of the SA systems. Since the development of ICTs turns city citizens to prosumers, such methods as Social Adaptive Case Management could be considered for providing Knowledge Management.

The PSAS could be used not only within SA program, but also for development and implementation of consumer identification systems. But toward to implementation of SC conception SA systems should fully support consumer identification, because supporting additional identification system resulted to additional effort to support necessary synchronization between these different systems. Governmental control, provision of high-level PSAS, integration of Information Systems successfully implemented in different organizations make it possible to avoid unnecessary efforts and extra complexity of urban infrastructure identification and inventory services.

It should be emphasized that new peculiarity named Aesthetics has been introduced. SA issues from point of view of ICTs and urban infrastructure/urban management are discussed in [1]. Analyzing the PSAS through a socio-cultural component of SD shows that Aesthetics should be introduced as the separate peculiarity.

ICTs and consequently SC are intensively developed technologies. This research used available SC literature that is not sufficient to confidently draw long-term conclusions. Urban development is an expensive sector of economy. The SA system can be evaluated in real life only if it is implemented city-wide and at least within the payback period of the SA program. All these limitations dictate necessity in further research in this topic. Directions of further research could be as follows:

- Formulation of variables and development of test procedure to expand this research;
- Development of SC oriented SA methodology;
- Development of methodology for assessment of the existing SA system and determination of needs in its upgrading;
- Evaluation of practical implementation of approaches discussed in this research;
- Research of cases where SC initiatives fail and connections of the fails with SA system (negative results should be studied deeply);
- Development of a Knowledge Management system, that could help accumulating the gained experience of research, implementation, and exploitation of SA programs in different cities;
- Development of training course (or upgrading existing ones) for civil engineers.

Undoubtedly, many of such researches are too expensive, but implementation of theoretical research at the planning stage can reduce cost of the SA program and increase its quality. It can be very attractive for investment planning as urban development/management/engineering as a whole. All these can affect urban living standards and SD.

Taking into account reductionism presented in modern science and its negative effect concluded in implementation of new technologies on business level without proper research, this research shows that synthesis of ICTs, civic science and economical science (the author means SD) can be very useful. SC being a complex living ecosystem for people requires complex and synthetic approaches in planning and management. As Bas Boorsma has written in his blog, technology myopia, solution-ism, no plan to replicate or scale, digital divides, lack of community communications, and closed architectures are ones of the hardships preventing SD of SC [31]. These hardships stress necessity to apply complex and synthetic approaches for SD of SC. So, this paper contributes science by adding new synthetic approach to analyze SA quality.

The uniqueness of the findings in the literature analysis of this study underlies in the fact that they belong to different fields and their synthesis allows us to draw interesting conclusions. Moreover, new findings are based on well-known ones. SD can be assessed on base of SA peculiarities, which in their turn can be assessed on base of well-known Software Quality "-ilities". Although there are no direct links between these conceptions in many cases and some conceptions introduced for the first time, presented approach describes the most understandable theory for both IT professionals and civil engineers.

Acknowledgement. I wish to express my sincere gratitude to SINAM Ltd. for their kind support and inspiration in the research.

References

1. Gakh, D.: Peculiarities of modern street addressing system toward to implementation of the smart city conception. In: Proceedings of the Federated Conference on Computer Science and Information Systems, pp. 543–552 (2020). https://doi.org/10.15439/2020F68
2. Tegmark, M.: An open letter: research priorities for robust and beneficial artificial intelligence. https://futureoflife.org/ai-open-letter. Accessed 9 Jan 2021
3. Gakh, D.: A review of street addressing systems within the realization of conception of smart city. In: Proceedings of the XII International Scientific-Practical Conference Internet-Education-Science IES-2020, Vinnytsia, pp. 96–98 (2020)
4. Farvacque-Vitkovic, C., Godin, L., Leroux, H., Verdet, F., Chavez, R.: Street Addressing and the Management of Cities. The WorldBank, Washington DC (2005)
5. Ziemba, E.: The ICT adoption in enterprises in the context of the sustainable information society. In: Proceedings of the Federated Conference on Computer Science and Information Systems, vol. 11, pp. 1031–1038 (2017). https://doi.org/10.15439/2017F89
6. Ziemba, E.: The ICT adoption in government units in the context of the sustainable information society. In: Proceedings of the Federated Conference on Computer Science and Information Systems, vol. 15, pp. 725–733 (2018). https://doi.org/10.15439/2018F116
7. Osuszek, Ł., Stanek, S.: The evolution of adaptive case management from a DSS and social collaboration perspective. In: Ziemba, E. (ed.) Information Technology for Management. LNBIP, vol. 243, pp. 3–16. Springer, Cham (2016). https://doi.org/10.1007/978-3-319-30528-8_1
8. Ziemba, E.: The holistic and systems approach to the sustainable information society. J. Comput. Inf. Syst. **54**(1), 106–116 (2013). https://doi.org/10.1080/08874417.2013.11645676
9. Mosannenzadeh, F., Vettorato, D.: Defining smart city. A conceptual framework based on keyword analysis. TeMA (2014)
10. Moustaka, V., Maitis, A., Vakali, A., Anthopoulos, L.: CityDNA dynamics: a model for smart city maturity and performance benchmarking. In: Companion Proceedings of the Web Conference 2020, WWW 2020, pp. 829–833 (2020). https://doi.org/10.1145/3366424.3386584
11. Kopackova, H., Libalova, P.: Smart city concept as socio-technical system. In: 2017 International Conference on Information and Digital Technologies (IDT), Zilina, pp. 198–205 (2017). https://doi.org/10.1109/DT.2017.8024297
12. Dharmavaram, S., Farvacque-Vitkovic, C.: Street addressing - a global trend. In: 2017 World Bank Conference on Land and Poverty. The World Bank, Washington DC (2017)
13. Centre for Civil Society: Do street vendors have a right to the city? IGLUS (2019)

14. Brad, J.: Examining the role of the construction industry in building smart cities. IGLUS (2020)
15. Walliman, N.: Research Methods. The Basics. Routledge, London (2011)
16. Delgado, J.: Service interoperability in the internet of things. In: Bessis, N., Xhafa, F., Varvarigou, D., Hill, R., Li, M. (eds.) Internet of Things and Inter-cooperative Computational Technologies for Collective Intelligence. SCI 2013, vol. 460, pp. 51–87. Springer, Heidelberg (2013). https://doi.org/10.1007/978-3-642-34952-2_3
17. Schauer, T.: The Sustainable Information Society – Vision and Risks. The Club of Rome – European Support Centre, Vienna (2003)
18. Blondelle, G., et al.: Polarsys towards long-term availability of engineering tools for embedded systems. In: Proceedings of the Sixth European Conference on Embedded Real Time Software and Systems (ERTS 2012), Toulouse, France, 1–2 February 2012. https://doi.org/10.13140/RG.2.1.4210.0325
19. Gamalielsson, J., Lundell, B.: Sustainability of Open Source software communities beyond a fork: how and why has the LibreOffice project evolved? J. Syst. Softw. **89**, 128–145 (2014)
20. Caprotti, F., Cowley, R., Flynn, A., Joss, S., Yu, L.: Smart-Eco cities in the UK: trends and city profiles. University of Exeter (SMART-ECO Project) (2016)
21. Anttiroiko, A.: City-as-a-platform: the rise of participatory innovation platforms in finnish cities. Sustainability **8**(9), 922 (2016). https://doi.org/10.3390/su8090922
22. Sumaray, A., Kami Makki, S.: A comparison of data serialization formats for optimal efficiency on a mobile platform. In: Proceedings of the 6th International Conference on Ubiquitous Information Management and Communication (ICUIMC 2012), Article no. 48, pp. 1–6 (2012)
23. Gakh, D.: Essentials of Street Addressing Programme for Kabul City. Ministry of Communications and Information Technology, Islamic Republic of Afghanistan, Kabul (2014)
24. Vugalter, M., Even, A.: The impact of enhancing a performance measurement system on the Israeli police. In: Ziemba, E. (ed.) Information Technology for Management. LNBIP, vol. 243, pp. 162–178. Springer, Cham (2016). https://doi.org/10.1007/978-3-319-30528-8_10
25. Chmielarz, W., Zborowski, M., Atasever, M.: On aspects of internet and mobile marketing from customer perspective. In: Ziemba, E. (ed.) AITM/ISM -2019. LNBIP, vol. 380, pp. 27–41. Springer, Cham (2020). https://doi.org/10.1007/978-3-030-43353-6_2
26. Alonazi, M., Beloff, N., White, M.: MGAUM—towards a mobile government adoption and utilization model: the case of Saudi Arabia. Int. J. Bus. Hum. Soc. Sci. **12**(3), 459–466 (2018)
27. Alonazi, M., Beloff, N., White, M.: Perceptions towards the adoption and utilization of M-government services: a study from the citizens' perspective in Saudi Arabia. In: Ziemba, E. (ed.) AITM/ISM -2019. LNBIP, vol. 380, pp. 3–26. Springer, Cham (2020). https://doi.org/10.1007/978-3-030-43353-6_1
28. van Dyk, R., Van Belle, J.-P.: Drivers and challenges for digital transformation in the South African retail industry. In: Ziemba, E. (ed.) AITM/ISM -2019. LNBIP, vol. 380, pp. 42–62. Springer, Cham (2020). https://doi.org/10.1007/978-3-030-43353-6_3
29. Kolasa, I., Papaj, T., Ziemba, E.: Information systems projects' success in government units: the issue of information systems integration. Procedia Comput. Sci. **176**, 2274–2286 (2020). https://doi.org/10.1016/j.procs.2020.09.286
30. Van der Alst, W.M.P., Berens, P.J.S.: Beyond workflow management: product-driven case handling. In: Ellis, S., Rodden, T., Zigurs, I. (eds.) International ACM SIG GROUP Conference on Supporting Group Work (GROUP 2001), pp. 42–51. ACM Press, New York (2001)
31. Boorsma, B.: When smart city initiatives fail - and why. 12 factors that contributed to smart city hardships, 27 February 2018. https://www.linkedin.com/pulse/when-smart-city-initiatives-fail-why-bas-boorsma/. Accessed 9 Jan 2021

Digital Transformation in Legal Metrology: An Approach to a Distributed Architecture for Consolidating Metrological Services and Data

Alexander Oppermann(✉), Samuel Eickelberg, and John Exner

Physikalisch-Technische Bundesanstalt, Abbestr. 2-12, 10587 Berlin, Germany
{alexander.oppermann,samuel.eickelberg,john.exner}@ptb.de
http://www.ptb.de

Abstract. In this paper the digital transformation of sovereign processes are outlined. The driving force is to streamline and innovate metrological processes for measuring instruments under legal control. The main challenge is to preserve the same level of trust and security for digitally transformed procedures as for conventional paper-based procedures. In the digital domain, a new balance has to be negotiated between security and process requirements without burdening the digital procedures with additional security prerequisites. Considering the strict legal framework, a distributed software architecture approach is presented that offers privacy, security and resilience. The cornerstone of the digital transformation is the Administrative Shell, representing measuring instruments in the digital domain. It will hold all measuring instrument related documents throughout its life cycle for the first time and offers access to all stakeholders. For this reason, a new security concept with additional roles is introduced including the Metrological Administrator. It manages digital metrological processes. At the same time, the platform approach seamlessly integrates existing public and private infrastructures. Moreover, a service hub is created with interdependent services that support the digital transformation of paper-based processes, such as verification and software update. These main use cases are introduced, and their requirements and implementation approach are outlined. At the current state of research, all evaluated metrological processes will still depend on human interaction. Currently, it is not possible to fully automatize and streamline the process chain to its full extent. Nevertheless, the presented results are important steps towards this long term goal.

Keywords: Digital transformation · Metrological processes · Administrative shell · Metrological administrator · Distributed software architecture · Weighing instruments · Legal Metrology

E. Ziemba and W. Chmielarz (Eds.): ISM 2020/FedCSIS-IST 2020, LNBIP 413, pp. 146–164, 2021.
https://doi.org/10.1007/978-3-030-71846-6_8

1 Introduction

AnGeWaNt[1] is a joint research project of six associates from different areas spanning across commercial partners in the weighing and construction industry, a Notified Body in Germany, such as the Physikalisch-Technische Bundesanstalt (PTB), a regional agency for innovation and European affairs, and the Institute for applied labor science (IfaA), responsible for aspects within socio-economic and human factors [1].

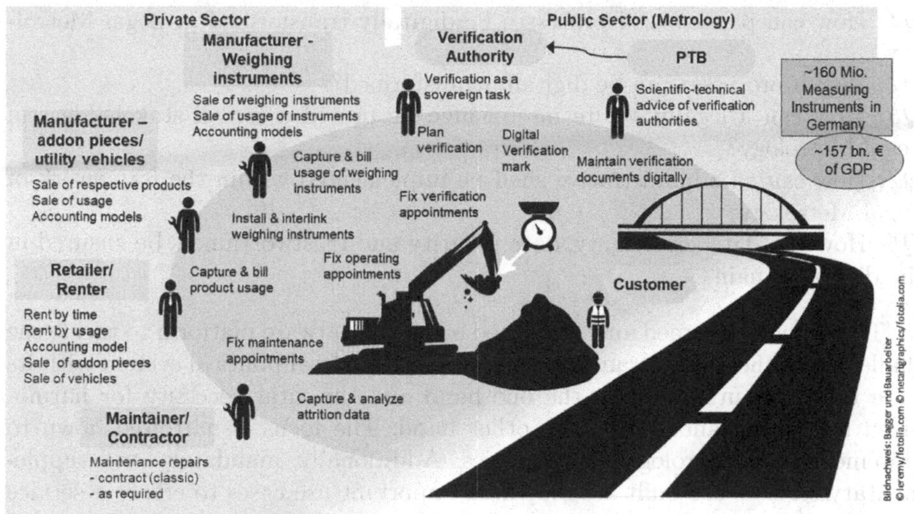

Fig. 1. Overview of the private and public sectors: The main stakeholders with their relevant tasks, and their future capabilities. Using weighing instruments in utility meters as the main use case for innovating legally relevant processes.

The field of Legal Metrology comprises around 160 million measuring instruments in Germany, which contribute about 157 billion Euros to the national GDP [2]. Its main purpose is to establish and provide trust among all stakeholders such as customers, manufacturers, and users of measuring instruments. Fostering trust in measuring instruments and their (remote) procedures is indispensable for a working economy. This holds true especially with regard to a growing technology stack and increasing complexity for applied instruments.

This paper is an extension to the prior published conference paper [3]. It focuses on innovating sovereign tasks and their digital transformation of paper-based procedures in Legal Metrology (Fig. 1). The two exemplary processes of verification and software update application have been designed using flowcharts focusing on interaction and data flow among different stakeholders. In addition,

[1] *Arbeit an geeichten Waagen für hybride Wiegeleistungen an Nutzfahrzeugen*, https://www.angewant.de/.

an approach is developed for measuring instruments under legal control that will require a uniquely identifying and descriptive digital representation. Therefore, the concept of a digital *Administrative Shell* has been devised by the authors. Moreover, a central role, the *Metrological Administrator*, is introduced. It governs digital processes in Legal Metrology. The Administrative Shell as well as the Metrological Administrator are cross-domain concepts, which require a common infrastructure and a unified process framework for future digital metrological processes. This is part of the research gap, which leads to the following research questions:

Q1: How can paper-based processes be digitally transformed in Legal Metrology?
Q2: Which processes can be digitally transformed?
Q3: How can a legally secure interchange be provided for all stakeholders in Legal Metrology?
Q4: How can an administrative shell be implemented within the framework of Legal Metrology?
Q5: How can data sovereignty, data security and trustworthiness be ensured in the digital domain?

The authors designed and developed a central software platform to tackle the challenges of the digital transformation within a federal political system with its wide range of authorities on the one hand and the vital necessity for harmonized technical standards on the other hand. The focus is narrowed down to aforementioned metrological procedures. Additionally, mandatory and supplementary services are built around these important use cases to create a service hub.

The remainder of this paper is structured as follows. Section 2 aligns the related work with the AnGeWaNt project. In Sect. 3, an overview of the research subjects and procedure is given, focusing on legal requirements, and use cases. Section 4 presents the research findings, matching the research subjects as a platform architecture approach. The conclusions, implications and limitations of this research work is summarized in Sect. 5.

2 Theoretical Background and Related Work

By law, a Notified Body such as the PTB in Germany, is obliged to carry out conformity assessments of measuring instruments. The essential requirements of the Measuring Instrument Directive (MID) [4], such as reproducibility, repeatability, durability and protection against corruption of measuring instruments and measurements, have to be fulfilled before entering the market. After the measuring instrument has been placed on the market, regulations for processes, such as re-verification, are not in the jurisdiction of the MID. Therefore, in Germany the Measurement and Verification Act (MessEG) [5] and the Measurement and Verification Enactment (MessEV) [6] are applied. Further requirements are deduced from the International Organization of Legal Metrology (OIML) and their guideline D31 [7], on a European level from WELMEC Guide 7.2 [8] and the WELMEC Guide 7.3 Reference Architectures [9] and finally on a national tier from the technical guidelines TR 031091 of the Federal Office for Information Security [10].

2.1 Related Work

European Metrology Cloud. The AnGeWaNt project can be considered as a national spin off of the *European Metrology Cloud Project* (EMC) focusing only on the weighing instruments. The EMC project circumcises 16 different measuring instrument classes in a supranational setting within a European legislation context. One of the major goals is to support the unified digital single market that the European Commission has issued. However, the digital transformation of Legal Metrology, the creation of one single platform that joins existing infrastructures and databases as well as to streamline metrological services [11] are congruent goals with the AnGeWaNt project. While the EMC project aims to distribute hardware nodes [12] that will provide essential parts of the envisioned infrastructure of the AnGeWaNt project can be hosted, split up and distributed on any *Platform as a Service provider* (PaaS). The European Metrology Cloud concept is also in line with the General Data Protection Regulation [13].

GAIA-X. The GAIA-X project started as a joint initiative of Germany and France in 2019 with the goal to build a sovereign digital European cloud ecosystem, that is efficient, secure and highly distributed. Major key features are data sovereignty, privacy by complying with the General Data Protection Regulation (GDPR), transparency and openness by supporting open-source principles and being flexible by building a modular and highly inter-operable platform for a broad spectrum of industry partners. These include Small and Medium Enterprises (SME) as well as government bodies [14].

The envisioned portfolio of services will enable a vivid ecosystem that connects all sectors of the economy and will accelerate the digital transformation of the European Union. The fundamental element of the GAIA-X ecosystem will be the GAIA-X node, that will offer interdependent services [15].

It will include, but is not limited to, a federated identity management with (external) identity data providers. The European Metrology Cloud project is aiming to be compatible with the GAIA-X interfaces. EMC is striving for a seamless integration into GAIA-X by taking advantage of the identity management and offer its services in the envisioned cloud ecosystem. Both projects are European initiatives that strive for the establishment of a digital quality infrastructure [16].

3 Research Subjects and Procedure

In the following sections the research subjects and approaches are provided to answer the previously outlined research questions. Section 3.1 addresses the

Administrative Shell. It introduces the envisioned concept and its gradual evolution into the final manifestation. Moreover, data sovereignty, security and trustworthiness are addressed in the digital domain.

Transforming paper-based processes into the digital domain will inevitably alter data and process handling. In Sect. 3.2 the approach of the Metrological Administrator is described, which is an exclusive role to supervise metrological processes and control the information flow of process data.

Sections 3.3 and 3.4 consider the digital transformation of the verification application and software update process. Both use cases are approaches for seamless integration and developing new remote procedures. Finally an attempt of securing integrity as well as trustworthiness is explained in Sect. 3.5.

3.1 Digital Administration Shell

The cornerstone of the digital transformation will be the *digital administration shell*. This concept will be developed gradually. It evolves from a *device pass*, that creates a unique mapping to a measuring instrument (Sect. 4.2). Based on this, the *digital type plate* will be created, that holds verification-specific information, e.g. the number of the EC type-examination certificate and the accuracy class of a scale. It will result in a general *document store* that hosts all documents related to a specific measuring instrument (Fig. 2). Moreover, the digital administration shell will lay the foundation for future services, like a a revision-safe archive store that will meet the legal archive obligations for measuring instruments throughout their life span.

The concept of the general document store is designed for a multi-tenant scenario. Furthermore, it will be implemented with a document based rights management. It has similarities to the UNIX file access rights management, so that reading, writing and sharing can be managed for each document differently and will not only depend on a role of a user.

Furthermore, five different types of relevant data have been classified, such as personal, telecommunication, metrological, measuring instrument specific and measurement data. Each of them are entangled by different directives and can have several distinct ownership that may have also have a temporal dimension. It is important to notice, that the owner of data might be entitled to manage access rights, yet authorized third parties are allowed by law to read certain information that does not belong to them but they are entitled to. This is one of many peculiarities to keep in mind for designing the digital administration shell.

Another important feature is to ensure traceability, in order to support legal compliance. Thus, existing data cannot be deleted. In case the data is changed, it will be versioned automatically. This envisioned fine grained access right management will allow all stakeholders in Legal Metrology to support the information flow across usual limitations without giving up control and security.

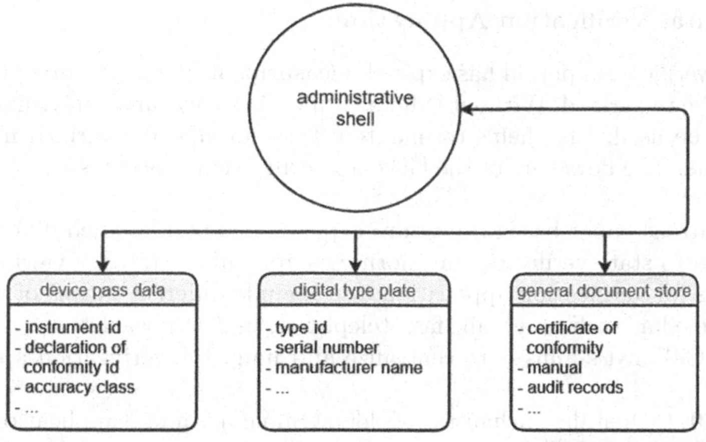

Fig. 2. Conceptual views of the administrative shell

The "Industrie 4.0 Plattform" published a recent concept of the digital twin and asset administration shell [17]. Furthermore, the platform issued the tool "AASX Explorer" to create administration shells for different business use cases. Within AnGeWaNt, this tool will be used as a template provider to implement a digital administration shell that will be compatible with the standard of the industry 4.0 consortium while fulfilling special requirements in Legal Metrology. The aim is to preserve a standardized data structure by extending it for measuring instruments under legal control.

3.2 Metrological Administrator

The metrological administrator is concerned with managing process data, while the administrative shell (Fig. 2) coordinates access to instrument-specific information. The metrological administrator has the rights to perform the following tasks:

- administration of metrological processes such as software update application,
- administration of German electronic file service (CRUD operations),
- administration of digital certificates (issue, revoke, invalidate),
- template provision for access rights for users,
- template provision for access rights for data fields,
- template provision for new digital certificates, and
- process supervision of German electronic file service.

The metrological administrator embodies the role of a process manager. The role exclusively grants control of the information flow in the metrological processes as well as the inter-process communication. Due to security concerns, the role has no direct access to the data generated in the administrative shell.

3.3 Digital Verification Application

After the verification period has expired, measuring instruments under legal control must be re-verified. Without valid a verification, measuring instruments may no longer be used. This helps to maintain trust in officially verified measuring instruments. The flowchart of the German verification process is made available [18].

The current paper-based procedure is performed and handled differently for each of the 13 state verification authorities across all 16 states of Germany. The procedures for verification application also include different means of transmission and media, such as eMail, fax, telephone, and written letters. Therefore, one of AnGeWaNt's aims is to centralize and unify the verification application process.

DEMOL ("Digitaler Eichantrag Melden Online") is a web application created by the verification authorities in Germany to unify the transmission of verification applications. It provides single or mass instrument verification applications via web interface. The AnGeWaNt platform connects to DEMOL via a specific REST interface (Sect. 4.2) to submit verification applications automatically.

3.4 Software Update Application

In Legal Metrology, software can be separated into a legally relevant and a legally non-relevant part. The legally relevant part is determined by its possible influence on the measurement result. If the user of a measuring instrument wants to update the legally relevant part, a software update application has to be submitted to the verification authorities.

The software update procedure of a measuring instrument is a complex process, which involves all stakeholders in Legal Metrology. Expert interviews provide insight about the administrative process flow that need to be integrated into digital transformation of the metrological process. As a result, flowcharts have been created which document and outline two scenarios: a standard appeal process [19] for regular updates, and an emergency appeal process [20] for critical security updates.

Software update applications are submitted by the user of measuring instruments but can be substituted by the manufacturer with a prior consent from the user. As a direct result of the digital transformation, the required data for the application, such as device-specific or applicant information, is provided by a central common data service of the AnGeWaNt platform (Sect. 4.2: Common Principal Data Service). In addition, the following information are required for a successful application: the type of the actual update procedure (physical media, wireless, or web download), the type approval, the operating manual, a written agreement of the user of the measuring instruments to update the software, a certificate of conformity (CoC) according to article 40 of MessEV, and finally, a written declaration that the same application has not yet been submitted by any other market and user surveillance.

When an application is submitted, it is checked for completeness and plausibility. A specified number of test devices with the software update, which will be tested in the field, is composed and the tests are carried out. If the tests were successful, the approval for a software update is granted for the applied measuring instruments, and the applicant is notified of the results. Finally, the applicant receives permission to further operate the updated measuring instruments.

3.5 Remote Authenticity

In the field, user and market surveillance verifies measuring instruments. With increasing usage of software in measuring instruments, a further decoupling of hardware and software is inevitable. As a consequence, only the sensor and a communication unit of the measuring instrument will remain in the field. Software will be outsourced into data centers ([21]). Thus, it is in the utmost interest of the authorities to verify legal relevant software parts remotely. Despite its location and physical access. Oppermann et al. [22] already addressed this problem and created a virtual verification monitor that fulfills the need of user and market surveillance. The authorities must have the possibility to remotely verify the authenticity of a used meter, processing unit, and associated logbook. The legally relevant software has to provide a form of identification. An acceptable solution according to the WELMEC Guide [8] would be: a software name consisting of a string of numbers, letters, or other characters, a string added by a version number, and a checksum over the code base.

Parts of this research concept [22] went into the development of the digital authenticity service (Sect. 4.2). This remote service can be used as a preliminary check to prepare the verification and calibration audit of the legal authorities. At the moment, the envisioned remote authenticity service will not be able replace a complete audit in the field, since the on-site manipulation possibilities still predominate. However, it can be used to frequently check the parameters of measuring instruments for possible changes, without the need to deploy additional staff into the field.

4 Research Findings

This section deals with the results of the previously outlined research subjects. The core finding is a highly decoupled platform to provide configuration flexibility, a unified user experience across all use cases, as well as strong information security.

The architecture of the AnGeWaNt platform is described in Sect. 4.1 as a central hub which integrates different back-end services into a common web front-end. The different back-end services are further elaborated in Sect. 4.2. Finally, security aspects as well as the rights and roles concept are handled in Sections 4.3 and 4.4.

4.1 Platform Architecture

The architecture of the platform is designed and implemented according to *SOA* (Service-oriented architecture). SOA is a software architecture model which classifies software components as services. These services are distinct units, stateless, loosely coupled and can be combined flexibly. The units communicate via *REST* (REpresentational State Transfer). REST requires HTTP- or HTTPS-based communication without the need of adding additional protocols (Sect. 4.1 Design Decisions).

As a central hub, the envisioned platform will offer access to all connected infrastructures and their provided services. All stakeholders can take advantage of the offered services and data within the platform domain. Consequently, this service hub is a first step to an interdependent service ecosystem. It will provide opportunities to develop new data-driven use cases beyond the current realm of stakeholders.

Furthermore, by its modular approach, the architecture allows new services to be added with little effort. To increase flexibility and ease later expandability, the project strives for standardized and harmonized interfaces across all services. The distributed architecture offers independent deployment as well as operation of services. The setup of the platform (Fig. 3) consists of three independent modules. The main module is the AnGeWaNt platform (upper left-hand dashed box). It offers a web-based user interface and services, such as verification application and software update. The user management module (lower left-hand dashed box) consists of the user and token manager services. That offers a secure, stateless and flexible authentication and authorization layer. Third-party systems are tied to AnGeWaNt in the external infrastructure module (right-hand dashed box), such as DEMOL or manufacturers' systems, to provide verification applications or device passes.

Design Decisions. Each service is implemented as its own application container, allowing for separate, independent deployment, maintenance, and operation. This fulfills the software separation requirement (S1) of WELMEC 7.2 Software Guide [8]. The main reason for this design decision is to operate each of the services independently, even across different environments. The second aspect is to be able to tailor the platform to specific configurations. As opposed to a monolithic architecture, each service can be updated independently, avoiding to update the whole platform at once.

Best practices of modern, distributed software development are applied according to [23] with regard to the software life cycle, ensuring maintainability and extensibility of the platform. Implementation is done in Java. It is the most common and well adopted language for client-server business application development[2]. *Spring Boot* is used as the base application framework. It provides pre-configured, managed dependencies to simplify build configurations.

[2] https://www.northeastern.edu/graduate/blog/most-popular-programming-languages/.

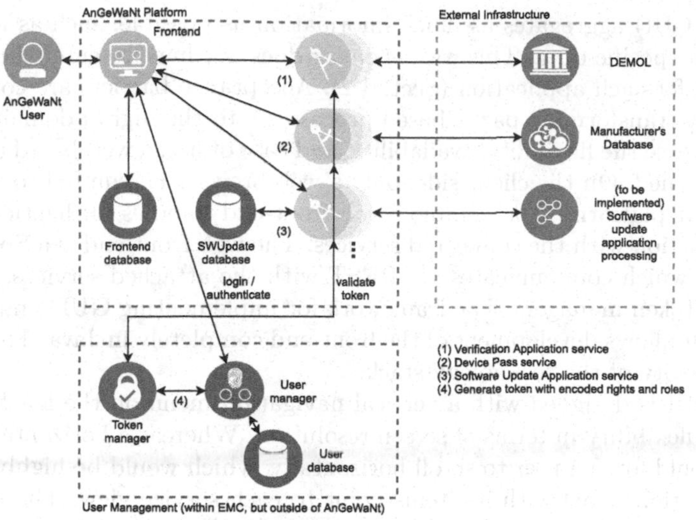

Fig. 3. Overview of the platform architecture concept

The required boilerplate code to get an application container up and running is already in place. The framework is lightweight and minimizes configuration effort to develop services[3].

Data Layer Abstraction. The underlying data structure is generated by Hibernate[4] using Java Persistence API entities (JPA)[5]. Data access is achieved by JPA repositories. They provide the standard CRUD operations by default. All services relying on persistence capabilities can be operated independently of any relational database management system. Without re-initializing the database, changes can be made to the data structure and are handled automatically.

User Experience. Another advantage of the AnGeWaNt approach is to design current processes from a user's perspective. This leads to additional benefits, such as plausibility checks, secure transmission and pre-filled forms. The user also gains transparency of the application processes by checking on the status of the current process.

Graphical User Interface. When designing user interaction, it is necessary to keep the user experience as intuitive as possible, while keeping domain-specific accuracy to the process logic. This holds true for application-relevant data, which can be used across different applications. The AnGeWaNt graphical user

[3] https://spring.io/projects/spring-boot.
[4] https://hibernate.org/.
[5] https://www.oracle.com/java/technologies/persistence-jsp.html.

interface (GUI) aggregates as much information as possible, such as measuring instrument specific data. This way, the user does not have to enter the information again for each application (Sect. 4.2). Acceptance barriers are consistently reduced by transforming paper-based processes into the digital domain.

To increase the flexibility, availability, and ease of use, a web-based user interface is designed. On the client side, only a web browser is required to access the AnGeWaNt platform. The common web front-end handles authentication and user interaction with the connected services. The web front-end is a Spring Boot container, which communicates via REST with the attached services, including User and Token managers. The framework for implementing GUI components is *Vaadin*[6]. It allows development of the front-end completely in Java. This reduces the complexity of the technology stack.

The GUI is designed with a vertical navigation menu on the left-hand side, providing flexibility in terms of screen resolution. Whereas a horizontal top navigation would force a user to scroll horizontally, which would be highly unusual. The navigation panel with its items point to each service view. The remainder of the screen is used to display the content panel.

4.2 Services

The process design has been done using flowcharts similar to Business Process Model Notation (BPMN), in order to visualize and document the processes. The charts describe each state in a process as well as their transitions and conditionals. They are based on expert interviews about the administrative workflows.

The interaction of stakeholders has been essential while designing the processes of software update application as well as verification application. Each stakeholder is located in its own swim lane, including information flows among them ([19], [20], and [18]). Deriving interfaces according to the modeled information flows has been a straight-forward task.

Each use case is implemented as a separate back-end service. They all provide REST endpoints for communication, such as triggering actions, or providing information. The following sections describe the most important services of the AnGeWaNt platform.

Digital Verification Application Service. This service provides an endpoint for submitting verification applications towards the external DEMOL system (Sect. 3). It checks plausibility of the application before submission. The service converts it to the specified XML structure and passes it on via the REST endpoint. This REST interface has been added as a finding of standardization effort. The service also reads the submission result from DEMOL, providing information on status of each of the instruments in the application and scheduled verification appointments.

[6] https://vaadin.com.

Software Update Application Service. The service main objective is to coordinate distribution of software updates to specified measuring instruments in the field. It has specified REST endpoints for submission of an application, updating status, receiving and placing a response to a hearing. These are the only steps in the application process that require interaction via the AnGeWaNt platform according to [20].

The software update application is processed according to the steps in Fig. 4. A new application is submitted, and the applicant as well as addressed authorities are being notified of the successful submission. Then the verification authority declares the test instruments for conducting the approval of the update. After specifying the test lot, additional information can be requested in a hearing of the applicant by the authority, which is necessary to conduct the approval. Afterwards, the application is either approved, or denied. If approved, the lot of instruments to be updated is being provisioned for carrying out the software update. Finally, the update results are being published.

Process transparency is being assured by validating status updates which can be triggered via REST endpoints. All issued applications can be viewed within the web front-end.

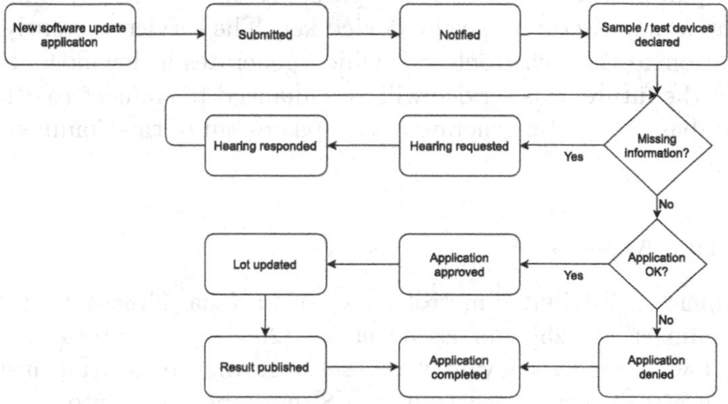

Fig. 4. Processing flowchart of the software update request process from the AnGeWaNt perspective

Digital Authenticity Service. Software, which controls and enables legally relevant procedures, must be able to verify itself for authenticity, which meets WELMEC 7.2 Software Guide [8] requirements U2 (Software Identification) and U8 (Software authenticity and Presentation of Results). That includes all services that perform such procedures, as well as the web front-end providing a GUI to these services ([22]). NIST "encourages application and protocol designers to implement SHA2-256 at a minimum for any applications of hash functions requiring interoperability" [24]. The calculated hash values of each service which performs legally relevant procedures as well as the AnGeWaNt front-end are stored separately in a *vault*[7]. This vault is exclusively accessed by the authenticity service. The service stores the hash values securely and verifies expected hash values against actual ones, without disclosure of the expected hash value.

Common Principal Data Service. Information from manufacturers, notified bodies, authorities, users, measuring instruments, and device types are used across different types of processes and documents. The common principal data service handles all data and is accessible through the front-end to fill in e.g. a software update application. This service is the essential component to facilitate the user experience in terms of data aggregation and intelligently pre-filling out forms. The API to access common principal data is designed according to the style guidelines[8] and harmonized across all types of data.

Device Pass Service. Device passes are the digital representation of physical device type plates, plus additional information, which may not fit on a physical type plate. The administrative shell (Sect. 3.1) of a measuring instrument uses the device pass as its instrument-specific information source. One of the partnering manufacturers operates its own database of device passes, from which the AnGeWaNt platform can retrieve device passes as JSON structured information by providing a manufacturer-specific device key. The service currently forwards the information to the web front-end, which generates a downloadable, secure PDF file. In the future, this service will be enhanced to connect to other manufacturer databases, and also generate device passes out of the Common Principal Data Service.

4.3 Security Aspects

Risk assessment of distributed metrological software has already been addressed by Oppermann et al. [25] and Esche et al. [26]. As the prototype platform is concerned with legally relevant processes regarding measuring instruments, the following attack vectors according to [8] must be taken into account when choosing an application framework: A_WEB_XSS (Cross-site scripting attack), A_WEB_DOS (Denial-of-Service attack), and A_WEB_SOCKET (introducing malicious code via web socket). Resilience against Cross-site scripting as well as Web socket security are supported by the employed framework. Denial-of-Service attacks are usually caught by the application server itself, not by the application framework. Authorized roles are mapped to specific paths and any other requests are being denied. The employed application framework addresses the aforementioned security concerns, given that the appropriate implementations and configurations are being carried out.

[7] A vault is an external secure store for secret information. The vault implementation used here is Spring Vault (https://spring.io/projects/spring-vault).

[8] Zalando RESTful API and Event Scheme Guidelines, https://opensource.zalando.com/restful-api-guidelines/.

4.4 Role, User and Device Management

The project aims to separate user and device credentials, as well as rights and roles management from the AnGeWaNt platform. First, the User manager and Token manager services can be re-used in other projects and services across the EMC. Second, all user-related data is decoupled from any specific application, increasing the security of the platform. This ensures the protective software interface requirement (Software Separation Requirement S3) outlined in the WELMEC 7.2 Software Guide [8].

Another design paradigm is to avoid sessions, because they cannot be held in a highly distributed architecture across potentially different domains. Instead, tokens are being generated and assigned to either a user or a device which authenticate themselves, or a service to prove its authenticity. All authentication and rights information of a user or a device is encoded in the token. It is a standard JWT (JSON Web Token), and it is generated after successful login by the Token manager service. It provides single sign on (SSO) for all authorized applications and services within the AnGeWaNt realm.

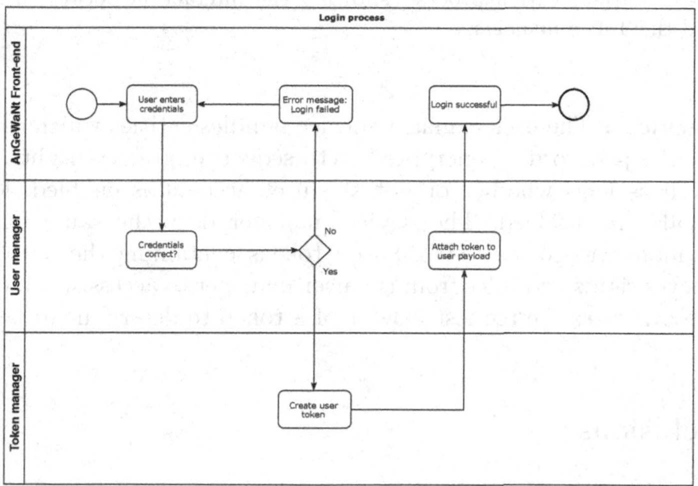

Fig. 5. User login process, showing the interaction between AnGeWaNt front-end, User and Token managers

Process Flow. The user authentication process (Fig. 5) deals with *authenticity* ("Am I a valid user?") and *authorization* ("Do I have the required rights/roles to access the resource?"). By communicating with stateless tokens, a session which contains currently authenticated user information is not required.

Any service requesting a provided token's authenticity from the token manager must provide the shared secret of the token manager in the request. This is necessary to ensure that a valid token can only be issued by the token manager.

The service authentication process (Fig. 6) is only concerned with authenticity ("Am I a valid service, that is, do I have a valid token?"), not authorization. In addition, a device authentication is handled the same way as a user authentication. The device manager holds the device's rights and its authentication token.

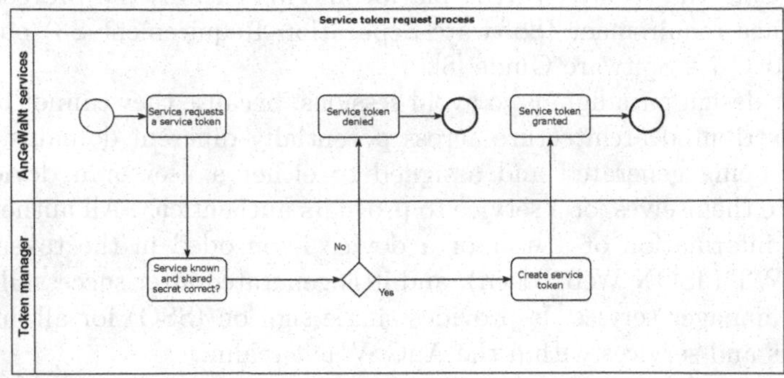

Fig. 6. Service authentication process, showing the interaction between AnGeWaNt services, and the Token manager

Access Services. The user manager service handles entities which contain the user name, the password as encrypted byte sequence, granted rights, assigned roles, as well as flags whether or not the user account is enabled, locked, or its credentials are disabled. The device manager does the same for devices. The token manager provides JWT-based tokens containing the authorization, which retrieves rights and roles from the user manager to access specific services. The service evaluates the request header of a token to determine an authorized request.

5 Conclusions

5.1 Research Contribution

In this paper, the AnGeWaNt project introduces its approach to seamlessly integrate existing public and private infrastructure. The legal framework with its security requirements are briefly described and also its implication for the platform design. Legally regulated processes, such as verification application and software update, are central use cases. Furthermore, the extensible concept of the administrative shell is introduced, which enables a central document store throughout the life span of the measuring instrument. Enabling an information flow across usual limitations for all stakeholders without surrendering trust and security. The introduced metrological administrator assures a process flow that is legally compliant. The digital transformation is the driving force to innovate and streamline the underlying infrastructure.

5.2 Implications for Research and Practice

By creating a central platform with a service hub, the digital transformation of paper-based procedures is supported. The platform is modular and built to integrate independent services. The administrative shell as a central document store streamlines the management and information flow of metrological processes. This provides new opportunities beyond the classical realm of all stakeholders and increases data-driven innovation for future success. Furthermore, each service can be easily separated and integrated into another context. This will assure that the progress achieved in this project can be reused and will avoid time intensive and costly reimplementations. The highly modularized nature of this platform allows a greater flexibility, adaptability and eases the burden of distributing the services across different domains. This increases indirectly the resilience of the platform.

Furthermore, the AnGeWaNt platform is designed from the beginning with *multi-tenancy* concept in mind. The need for different stakeholders, such as manufacturers, market and user surveillance authorities, and notified bodies to access only its own relevant data is crucial. This is guaranteed by the *User Manager*. With the introduction of the *Token Manager*, a compact, single sign on solution is created that is used to verify a user and grant access to documents across infrastructures. This is especially helpful, e.g. to send verification applications to external systems like DEMOL or to import device passes from manufacturers. Moreover, it implements a sessionless authentication concept for distributed systems across different domains.

5.3 Limitations and Future Work

While analysing and documenting metrological processes, it became obvious that not all steps of the process chain can be digitally transformed. At the moment, human interaction will always be necessary to carry out required steps in the field. This is an apparent limitation for all evaluated services, that will be extended by a remote capacity. Another limitation is concerned with issuing digital certificates to grant operation of measuring instruments after e.g. a software update. A standardized certificate as a digital, secure representation does not yet exist and is not within scope of the AnGeWaNt project.

In the next iteration, *multi-factor authentication* will be introduced, to increase security and enhance trust in the platform. The user manager will be extended to use a second factor for authentication. This can be a mobile device running an authenticator app, or a physical device, e.g. a fingerprint scanner attached to a client computer. Furthermore, the support and integration of OpenID connect framework[9] is intended. In order to harden the platform prototype, a risk assessment especially designed for distributed systems will be carried out.

---

[9] https://openid.net/connect/.

The *Guide to the Expression of Uncertainty in Measurement* (GUM)[10] is a metrological-mathematical collection of rules for the determination and denomination of measurement uncertainties [27]. In a future iteration of the prototype, an uncertainty analysis service can be implemented to e.g. run analyses of verification results of a measuring instrument. Hall elaborates causes of measurement traceability using standard methods of uncertainty propagation [28]. At PTB, the mathematical models behind GUM have been translated into a set of *Mathematica*[11] scripts, which in turn produce generic XML files for testing actual uncertainty calculations against GUM specifications ([27]). The first objectives will be to evaluate how these XML files may be used within such a service, and then how the uncertainty propagation can lead to useful insights of analyzing verification results.

In the long run, the *Digital Administration Shell* will be extended to import an XML-based Certificate of Conformity in the near future. This will offer new services, e.g. remote verification, issuing and revoking certificates. The PTB is working to establish the same level of trust for digital certificates as conventional paper-based certificates guarantee nowadays.

References

1. Ottersböck, N., Frost, M., Jeske, T., Hartmann, V.: Systematischer Kompetenzaufbau als Erfolgsfaktor zur Etablierung hybrider Geschäftsmodelle. In: Digitale Arbeit, digitaler Wandel, digitaler Mensch - 66. Kongress der Gesellschaft für Arbeitswissenschaft. GfA (2020)
2. Leffler, N., Thiel, F.: Im Geschäftsverkehr das richtige Maß - Das neue Mess und Eichgesetz, Schlaglichter der Wirtschaftspolitik (2013)
3. Oppermann, A., Eickelberg, S., Exner, J.: Toward digital transformation of processes in legal metrology for weighing instruments. In: 2020 15th Conference on Computer Science and Information Systems (FedCSIS). IEEE (2020). https://doi.org/10.15439/2020F77
4. European Parliament and Council: Directive 2014/32/EU of the European Parliament and of the Council. Official Journal of the European Union (2014)
5. Federal Ministry of Justice and Consumer Protection (BJFV): Gesetz über das Inverkehrbringen und die Bereitstellung von Messgeräten auf dem Markt, ihre Verwendung und Eichung sowie über Fertigpackungen (Mess- und Eichgesetz - MessEG) (2019). https://www.gesetze-im-internet.de/messeg/
6. Federal Ministry of Justice and Consumer Protection (BJFV): Verordnung über das Inverkehrbringen und die Bereitstellung von Messgeräten auf dem Markt sowie über ihre Verwendung und Eichung (Mess- und Eichverordnung - MessEV) (2020). https://www.gesetze-im-internet.de/messev/
7. Organisation Internationale de Métrologie Légale: General requirements for software controlled measuring instruments (2008)
8. WELMEC Committee: WELMEC 7.2 Software Guide. WELMEC European cooperation in legal metrology, Welmec Secretariat, Delft, Standard (2019)

[10] https://www.bipm.org/en/publications/guides/gum.html.

[11] Wolfram Mathematica, https://www.wolfram.com/mathematica/.

9. WELMEC Committee: WELMEC 7.3 Guide Reference Architectures - Based on WELMEC Guide 7.2. WELMEC European cooperation in legal metrology, Welmec Secretariat, Delft, Standard (2019)
10. BSI: Technische Richtlinie BSI TR-03109-1 Anforderungen an die Interoperabilität der Kommunikationseinheit eines intelligenten Messsystems. Bundesamt für Sicherheit in der Informationstechnik, Bonn (2013)
11. Thiel, F.: Digital transformation of legal metrology - the European metrology cloud. OIML Bull. **59**(1), 10–21 (2018)
12. Dohlus, M., Nischwitz, M., Yurchenko, A., Meyer, R., Wetzlich, J., Thiel, F.: Designing the European metrology cloud. OIML Bull. **61**(1), 08–17 (2020)
13. Thiel, F., Wetzlich, J.: The European metrology cloud: impact of european regulations on data protection and the free flow of non-personal data. In: Array (ed.) International Congress of Metrology, p. 01001 (2019). https://doi.org/10.1051/metrology/201901001
14. Federal Ministry for Economic Affairs and Energy (BMWi): Project GAIA-X - A Federated Data Infrastructure as the Cradle of a Vibrant European Ecosystem - Executive Summary. Official Journal of Federal Ministry for Economic Affairs and Energy (2019)
15. Federal Ministry for Economic Affairs and Energy (BMWi): Project GAIA-X - A Federated Data Infrastructure as the Cradle of a Vibrant European Ecosystem. Official Journal of Federal Ministry for Economic Affairs and Energy (2019)
16. Thiel, F., Nordholz, J.: Quality Infrastructure 'Digital' (QI-Digital). Federal Ministry for Economic Affairs and Energy (2020). https://www.bmwi.de/Redaktion/EN/Artikel/Digital-World/GAIA-X-Use-Cases/quality-infrastructure-digital-qi-digital.html
17. Boss, B., et al.: Digital twin and asset administration shell concepts and application in the industrial internet and industrie 4.0 (2020). https://www.plattform-i40.de/PI40/Redaktion/DE/Downloads/Publikation/Digital-Twin-and-Asset-Administration-Shell-Concepts.html
18. Exner, J., Oppermann, A.: German verification process (2019). https://www.angewant.de/wp-content/uploads/2020/06/Eichantrag.pdf
19. Exner, J., Oppermann, A.: German software update emergency appeal (2019). https://www.angewant.de/wp-content/uploads/2020/06/Standardverfahren_Softwareaktualisierung.pdf
20. Exner, J., Oppermann, A.: German software update process (2019). https://www.angewant.de/wp-content/uploads/2020/06/Eilverfahren_Softwareaktualisierung.pdf
21. Oppermann, A., Yurchenko, A., Esche, M., Seifert, J.P.: Secure cloud computing: multithreaded fully homomorphic encryption for legal metrology. In: Traore, I., Woungang, I., Awad, A. (eds.) ISDDC 2017. LNCS, vol. 10618, pp. 35–54. Springer, Cham (2017). https://doi.org/10.1007/978-3-319-69155-8_3
22. Oppermann, A., Toro, F.G., Thiel, F., Seifert, J.P.: Secure cloud computing: reference architecture for measuring instrument under legal control. Secur. Priv. **1**(3), e18 (2018). https://doi.org/10.1002/spy2.18
23. Martin, R.C.: Clean Code: A Handbook of Agile Software Craftsmanship. Prentice Hall, Upper Saddle River (2009)
24. Dworkin, M.: NIST Policy on Hash Functions (2015). https://csrc.nist.gov/Projects/Hash-Functions/NIST-Policy-on-Hash-Functions

25. Oppermann, A., Esche, M., Thiel, F., Seifert, J.P.: Secure cloud computing: risk analysis for secure cloud reference architecture in legal metrology. In: Accepted in Federated Conference on Computer Science and Information Systems (FedCSIS). IEEE (2108). https://doi.org/10.15439/2018F226
26. Esche, M., Thiel, F.: Software risk assessment for measuring instruments in legal metrology. In: Proceedings of the Federated Conference on Computer Science and Information Systems, pp. 1113–1123 (2015). https://doi.org/10.15439/2015F127
27. Greif, N., Schrepf, H.: A test environment for GUM conformity tests. In: PTB-Bericht. No. 19 in PTB-IT, Wirtschaftsverlag NW, Bremerhaven (2015)
28. Hall, B.D.: An opportunity to enhance the value of metrological traceability in digital systems. In: 2019 II Workshop on Metrology for Industry 4.0 and IoT (MetroInd4.0 IoT). Measurement Standards Laboratory of New Zealand (2020). https://doi.org/10.1109/METROI4.2019.8792841

Author Index

Agafonov, Anton 87

Bicevska, Zane 25
Bicevskis, Janis 25

Carchiolo, Vincenza 67

Delalic, Sead 103
Donko, Dzenana 103

Eickelberg, Samuel 146
Exner, John 146

Gakh, Dmitriy 127

Harada, Fumiko 45

Ito, Hiroki 45

Longheu, Alessandro 67

Malgeri, Michele 67
Mangioni, Giuseppe 67
Miller, Gloria J. 3
Myasnikov, Vladislav 87

Nikiforova, Anastasija 25

Oditis, Ivo 25
Oppermann, Alexander 146

Shimakawa, Hiromitsu 45
Supic, Haris 103

Trapani, Natalia 67

Yumaganov, Alexander 87

Zunic, Emir 103

Printed in the United States
by Baker & Taylor Publisher Services

Printed in the United States
by Baker & Taylor Publisher Services